T0073640

OUR
MOON

How Earth's Celestial Companion
Transformed the Planet, Guided Evolution,
and Made Us Who We Are

RANDOM HOUSE  NEW YORK

OUR MOON

Rebecca Boyle

Published in the United States by Random House, an imprint and division of Penguin Random House LLC, New York.

RANDOM HOUSE and the HOUSE colophon are registered trademarks of Penguin Random House LLC.

LIBRARY OF CONGRESS CATALOGING-IN-PUBLICATION DATA
Names: Boyle, Rebecca (Rebecca B.), author.
Title: Our moon : how Earth's celestial companion transformed the planet, guided evolution, and made us who we are / Rebecca Boyle.
Description: First edition. | New York : Random House, 2024 | Includes bibliographical references and index.
Identifiers: LCCN 2023011947 (print) | LCCN 2023011948 (ebook) | ISBN 9780593129722 (hardcover) | ISBN 9780593129739 (ebook)
Subjects: LCSH: Moon—Popular works. | Moon—History—Popular works. | Moon—Social aspects—Popular works. | Moon—Religious aspects—Popular works. | Moon—Philosophy—Popular works.
Classification: LCC QB581.9 .B695 2023 (print) | LCC QB581.9 (ebook) | DDC 523.3—dc23/eng/20231026
LC record available at https://lccn.loc.gov/2023011947
LC ebook record available at https://lccn.loc.gov/2023011948

Printed in the United States of America on acid-free paper

randomhousebooks.com

4 6 8 9 7 5 3

Book design by Simon M. Sullivan

For my daughters

Contents

The wonder of everyone is vanquished by the last star, the one most familiar to the Earth, and devised by nature to serve as a remedy for the shadows of darkness—The Moon.

—Pliny the Elder, *Natural History*

THE MOON WAS THE KEY.

Lieutenant Colonel David Shoup, a red-cheeked, thirty-nine-year-old marine, worried about the Moon as his boat sped through the black Pacific night. The Moon was not visible yet; at last quarter, waning away, it would rise over the ocean at midnight. But even an unseen Moon exerts a powerful influence on Earth.

Shoup knew the marines' boats needed at least four feet of water to float over the coral reef of the Tarawa Atoll, a hollow triangle of an island in the South Pacific Ocean. Today, Tarawa is the capital of the nation of Kiribati, but on November 20, 1943, it was step one in the Allies' plan to defeat the Empire of Japan. The capture of the tiny island and its airstrip depended on the tide, which depended on the Moon.

Allied battle planners scheduled their invasion for the morning of November 20, when they expected the high tide to crest over the reef surrounding Tarawa. Without satellite measurements, planners could only guess how much the tide would swell that day. They checked the lunar cycle against century-old tide charts from the Pacific, which was all they had, supplementing these with more recent data from places as distant as Australia and Chile. Planners estimated high tide would reach five feet by 11:15 A.M.—high enough for the ships with room to spare. But to Shoup, this depth was still too close for comfort.

"We'll either have to wade in with machine guns maybe shooting at us," he told *Life* correspondent Robert Sherrod, who was embedded with the troops, "or the [amphibious tractors] will have to run a shuttle service between the beach and the end of the shelf. We have got to calculate high tide pretty closely for the Higgins boats to make it."[1]

AS THE COMMANDERS fretted in the hours before sunrise, my grandfather Private First Class John J. Corcoran bent to his task on Nanomea Island, five hundred miles southeast. He was a small but important part of the largest armada in the Pacific Theater during World War II.

Jack, like so many young Americans in the middle of the war, enlisted eagerly and was ready to fight, newly equipped with a rifle, a bayonet, and a Marine Corps paycheck. He earned $6.40 per month and gained the skills to outfit airplanes with bombs.

In September 1943, at seventeen, instead of moving away to college, Jack boarded the transport ship *Puebla* and sailed west from San Diego, across the deep and beguiling Pacific, whose waters were nothing like the gray Atlantic he knew so well. By November 1943, he had gone farther from home than anyone he had ever known, including his Irish immigrant parents.

THE TROOPS THAT amassed to take Tarawa vastly outnumbered the Japanese force that attacked Pearl Harbor, as well as the Allied force that had spent a sweltering, muddy six months securing the island of Guadalcanal. The Marine Corps heading for the atoll supplied cover for the Allies' ships, and General Julian Smith promised the Navy would deliver "the greatest concentration of aerial bombardment and naval gunfire in the history of warfare."[2] My grandfather contributed to the nine hundred tons of ordnance the Allies would drop during the Battle of Tarawa, clearing the field for the invasion.

Despite all this, the fight for Tarawa would end in the worst casualties in such a short battle in the history of the Marine Corps. Of the 5,000 men who stormed the beach, 1,115 were killed and nearly 2,300 were wounded in just seventy-six hours of fighting.

The tide did not rise to five feet on the morning of November 20. It barely rose at all, and the transport ships couldn't get past the reef, just as Shoup had feared. At 6:48 A.M. local time, Smith and an admiral radioed a pilot observing from a float plane and asked, "Is reef covered with water?" "Negative," came the reply. Instead, the marines had to clamber out of their beached boats and wade across six hundred yards of water to shore, lifting their rifles over their heads. Facing relentless fire from Japanese forces, hundreds of marines were shot and some drowned in the high water surrounding the reef. It would be decades before anyone understood why the blood-dimmed tide didn't rise that day.

I GREW UP hearing only a couple of stories from my grandfather's service in World War II; like many veterans, he didn't like to talk about it. Through the National Archives, I learned that his unit, Marine Aircraft Group-31, island-hopped behind the men on the front lines, moving to conquered lands one at a time to set up bombing runs for the next phase of battle in the Pacific. Through my mom, I learned that Jack wasn't able to fall asleep in his tent-covered foxholes, and that he prayed the rosary despite his terror on island after island. A devout Catholic, he taught the words to his fellow marines. *Hail Mary, full of grace,* he repeated, trying to drown out cries of "Americans will die!" in accented English. Stories of these Japanese threats sailing like ghosts on the night wind chilled me to the bone when I was a kid. I wish I had asked more about them. And I wish I got to tell Jack, before he died in 2010, that he had the Moon to blame for the marines' losses at Tarawa.

○ ˙

EVERY DAY, ON every coastline on Earth, the tide changes the threshold where the land meets the sea. Boats in a harbor rise and fall against their docks as the day wears on. Beaches widen and narrow, and kelp, shells, or other ocean dregs left by receding tidewaters dry in the sand, far from lapping waves.

The tide ebbs and flows because of the Moon's gravity and, to a lesser extent, the Sun's. As the Moon moves around Earth, the two bodies tug on each other. The side of Earth that is closest to the Moon feels the tug a little more strongly, and the Moon pulls water toward itself, creating two bulges* in the world's oceans. The bulges create the high tide, which originates in the ocean and progresses toward the coasts. Twice a month, the Sun adds some tidal heft, as well. When it is lined up with the Moon, causing either a full Moon or an invisible new Moon, the Sun's gravity amplifies the bulging effect. This forms what are known as spring tides, and these bring higher high tides and lower low tides.

Seven days later, when the Moon is not lined up with the Sun but set apart at a ninety-degree angle, it looks half full.[3] We call this first quarter or last quarter. The Sun's gravity has less of an impact on the tides and produces what is called a neap tide. The high and low tides are less extreme at this point in the month.

Earth's geography also plays a role in how the water comes in. The continents change the tide's flow, and the depth of a shoreline changes how quickly the tide will rise or how slowly it will fall. And the Moon's location in orbit around Earth changes its gravitational pull, too. The Moon, like all celestial bodies, does not travel in a circle but in an ellipse, something that, as we'll see later, we learned from a seventeenth-century German astronomer obsessed with the Moon. The point on its orbit where it is most distant from Earth is known as apogee, and the point where it is closest is known as perigee. Three or four times a year the perigee coincides with a full Moon, which astrologers in the first part of the twenty-first century c.e. dubbed a "supermoon." The closer Moon yields exceptionally high and low tides. The more distant, apogee Moon—call it a micromoon—is slightly smaller in the night sky and has a weaker pull. But even a faraway Moon exerts a powerful influence over Earth.

THE MARINES PLANNED their invasion during a neap tide and couldn't understand why the tide not only failed to rise enough but did not rise at all, for almost two days. The "dodging tide," as war chroniclers later

---

* Gravity causes the bulge nearest the Moon. Forces including gravity and the centrifugal force form a bulge on the side farthest from the Moon, too, making two bulges. This equates to two high and two low tides a day, nearly everywhere on Earth as it spins (there are a few exceptions that experience one tidal cycle, like in the Gulf of Mexico, because sometimes Earth's continents get in the way). The time between low and high tide at a given location is six hours and twelve minutes.

called it, lay low over the Tarawa reef because the Moon was at apogee, its pull weak because it was so far from Earth. November 20 was one of only two days in 1943 to experience an apogean neap tide. Before the satellite era, and certainly before the marines captured the island and measured its geography, there was no way for American military planners to have known how dramatically this lunar alignment would affect the tides in Tarawa.

DESPITE THE CARNAGE, marines kept coming ashore, and the bombs kept falling. After three days of fighting, the waters finally returned and the marines took the atoll, but the devastation was complete. Americans back home were outraged, wondering how capturing such a tiny island could have led to such casualties.

MY GRANDFATHER'S UNIT arrived at Tarawa on New Year's Eve 1943. By then, the Allies controlled the island and the Navy Seabees had cleared the beaches of the bodies and the palm trees. Private First Class Corcoran continued his work, equipping planes with bombs for the next stage of the multipart Pacific plan. The Moon was four days old when he arrived at the battered atoll. It hung in the evening sky like a scimitar, like a scythe, like the horns of a bull. It was small enough that you could easily miss it, until it snuck up on you.

$$\bigcirc \; {}^{\bullet}$$

AFTER TARAWA'S BLOODBATH, the Allies would pay more attention to the Moon's influence. It played a seminal role in the long months leading up to the invasion of Normandy and the liberation of France. The French Resistance relied on full Moons to safely drop spies and supplies by parachute, and the Allies knew they would need a full Moon's high tide and its glittering light to land their ships on the shores of mainland Europe and wrest it from the grip of Nazi Germany. In 1943, as Allied planners deliberated on the right port, the hopes of the French waxed and waned with every twenty-eight-day cycle that passed.

Finally, the Allies decided to invade at Normandy, on the northern coast of France, because it was close to the British coast but less obvious than Calais, a larger port city. Normandy had both a small port and a small airfield, which the Allies figured they could capture on the first day.[4]

FIRST THE ALLIES would cross the English Channel at night. Paratroopers would glide across, using only moonlight as a guide, aiming to capture two bridges. Then the heavy planes would take off. Commanders wanted about forty minutes of morning light to bomb the coast, just like at Tarawa, before the land invasion began. The ground assault troops would sail in on a low but quickly rising morning tide.[5]

On the Calvados coast of Normandy, the English Channel can rise by an astonishing nineteen feet from low to high tide. As at Tarawa, the change in the tidal zone is caused by the Moon's interaction with the shore's geography. Steep shores like those in Normandy can increase an intertidal zone by many times the height predicted by the rise and fall of water.*

The dramatic change in tide means the water flows in fast—on D-Day, it rose at a rate of a foot every fifteen minutes. The Allies figured about half an hour at low tide was enough for the first wave of forces to clear the beaches of mines, triangular wooden obstacles, and human-sized iron barriers installed by Nazi troops. If the tide then came in quickly, it would float Allied forces over the sands and into France as the invasion progressed.

In Normandy, a low tide near sunrise only occurs during either a new Moon or a full Moon. The airborne divisions needed the light of the full Moon to fly by, so it was settled: The Moon was the key once more. Winston Churchill recalled in his memoirs that the Allies pinpointed June 5, 6, and 7 for the invasion. "Only on three days in each lunar month were all the desired conditions fulfilled," he wrote. "If the weather were not propitious on any of those three days, the whole operation would have to be postponed at least a fortnight—indeed, a whole month if we waited for the Moon."[6]

---

* Two millennia earlier, another military leader named Julius Caesar—"the noblest man that ever lived in the tide of times," as Shakespeare called him—would learn about extreme Channel tides the hard way when his first attempt to invade Britain failed.

THE FORECAST FOR June 5 was poor and Supreme Commander Dwight D. Eisenhower postponed D-Day by a day. The forecast was clear the next morning, and D-Day was a go for June 6. A full Moon rose on the evening of June 5 an hour and a half before sunset, reaching its highest point by 11:30 P.M. Greenwich Mean Time. At midnight, June 6, 1944 C.E., paratroopers from the 82nd and 101st Airborne divisions began dropping from the sky over France. Brigadier General James Gavin, from the 82nd Airborne, recalled that he could see clearly by the light of the Moon. "The roads and the small clusters of houses in the Normandy villages stood out sharply in the moonlight," he recalled.[7] The paratroopers took over two bridges to block Nazi tanks, and in the fields below, Resistance fighters bicycled by moonlight to cut off railroad tracks, subterranean telephone lines, and aerial power lines. Normandy was isolated on all sides. The first waves of troops crashed onto the beach at 6:30 A.M., under a morning summer Sun and a setting Moon.

The fighting lasted all day and the Allies suffered huge losses, especially at Omaha Beach. But by nightfall, the German forces were in retreat. After D-Day, the Allies liberated Paris and marched eastward to Berlin through the fall and winter, surviving the Germans' last-ditch offensive in the Battle of the Bulge. Germany surrendered on May 8, 1945.

My grandfather was in Havelock, North Carolina, that spring preparing for a land invasion of his own. The worst-case scenario, an invasion of Japan, required the marines to remain battle ready. Jack Corcoran was in training camp when the Nazis surrendered and he was still there in August 1945, when the United States dropped two atomic bombs on the cities of Hiroshima and Nagasaki. The Empire of Japan surrendered, and two months later Jack Corcoran was honorably discharged.

He went home to New Jersey. He married my grandmother, Helen, had six children and ten grandchildren, and retired after a long career as an accountant. Jack took me to the ocean every time I visited him in Toms River. I would stand in the sand of Seaside Heights with my brother and wait, watching, as the tide rolled in to reach me.

○˙

THESE TIDES AND these battles are just one part of the story of our Moon and ourselves. The Moon's role in the second great war is just a microcosm of the journey we have taken with it since our species emerged.

The Moon has shaped our rulers, and their conquests, since civilization's earliest days, but its power over us is far more ancient than even our conflicts. Its influence goes back to the sulfurous origins of this planet and everything that crawls, flaps, swims, or strains skyward on its surface. The Moon guides all of us from its vaulted position above us. But it's not apart from us, not least because it is actually a part of Earth. It was sheared from Earth when the planet was still freshly baked. And its elliptical orbit does not technically circle Earth, at least not in the way you might think. Instead, Earth and the Moon orbit each other, pivoting around a combined center of gravity that guides them both and that shapes their shared history.

Today, the Moon directs migrations, reproduction, the movements of the leaves of plants, and possibly the very blood in your veins. The Moon conducts the symphony of life on Earth, from the people who wage war on one another to the coral polyps that built the reefs of Tarawa. It has guided evolution since the moment of life's first stirrings, which occurred either inside deep ocean vents or in warm little pools at the water's edge, both of which derive nutrients through the Moon's tide.

The Moon makes Earth unique, certainly in our solar system and possibly in the broader cosmos. It made us who we are, in ways that scientists are just beginning to understand, from our physiology to our psychology. It taught us how to tell time, which we used to impose order on the world. The Moon inspired the human projects of religion, philosophy, science, and discovery.

This book is the story of our journey with the Moon in three parts: how the Moon was made, how the Moon made us, and how we made the Moon in our image. This is not solely an astronomy book, and it is not an Apollo book, though astronomy and the Apollo missions are both inseparable from humanity's journey with the Moon. This is a book about time, life on Earth, human civilization, our place in the universe, and how the Moon has made all of it possible. I hope this book changes your understanding of all these things. And I hope it changes the way you see the Moon, this partner world that has always been with you, and which I hope you notice anew the next time you go outside at night.

PART I

HOW THE MOON
WAS MADE

CHAPTER ONE

A World Apart

THE MOON IS DIFFERENT.

It is like nowhere on Earth, which is a watery bubble improbably bursting with life in a universe of emptiness. The Moon is barren and has been throughout the four-and-a-half-billion-year eternity of its companionship with this planet. The Moon is silent. It plays host to no cricket chorus, coyote calls, or night wind sailing through pines. It is dry, at least on the outside. There are no waves lapping on shores, no soft rains, no snow. It is a crater-pocked wasteland that smells of doused firecrackers. The Moon is scorching hot during its long day, and freezing cold during its long night.

The lunar landscape is grayscale, but flecked with shades of tan, chocolate, beach sand, chalk, gold, spicy-mustard ochre, and, in the words of Apollo 11 astronaut Michael Collins, a "cheery rose" hue.[1]

Sunlight on the airless Moon plays tricks on human eyesight, warping a moonwalker's sense of crater depths and hillside angles, making tiny slopes look like vertiginous peaks. All is monotony. There is no blue, and there is no green. No sunlight scatters through a watery atmosphere. No lichens splotch the Moon's rocks. No bacteria grow in its dirt to help plants flourish. Certainly, there are no birds overhead, ants underfoot, or any other kind of animal anywhere. On the Moon there is nothing and no one. Until the Apollo landings, no creatures ever looked up at the Moon's black sky and wondered about their place in it all. No one ever stared up at the crescent Earth and thought about visiting. There is no culture, except the one we brought.

The Moon says nothing for itself, but it says plenty about us. We project our dreams and our fervor onto its mottled surface and it serves as a mirror, both figuratively and literally. It reflects sunlight and even Earth's own light, ashen earthshine, back to us. We can see this phenomenon when the Moon is a crescent, and yet its full disk is just barely perceptible. The Moon is Earth in inverse, a desolate rock whose scars whisper of our world's violent past and underscore its riotous gardens of color and life. The Moon contains only what we imagine it to contain. It harbors only what we berth in its seas.

SINCE THE BEGINNING of time, the Moon has controlled life on Earth and shepherded the human mind through a spectacular journey of thought, wonder, power, knowledge, and myth. But this frenzied, multifarious, Earthly history disguises the truth of the Moon. As vivid and lively as our history with it has been, the Moon itself is quiet, barren, and still.

This was not always the case: When the Moon was young, it was livid with energy and heat, a magnetic field, oceans of lava, and maybe an active crust like the one that warps and wrinkles the face of our world. But no one was around for this lively phase. The only Moon we have ever known is the spectral one in our sky, the two-dimensional one, the cold and silent one.

Nothing happens there, except the occasional arrival of an asteroid or the briefly violent puff of a crashed or landed spacecraft. Nothing looks up, nothing breathes, nothing hopes.

When Apollo 11 astronaut Buzz Aldrin walked on the Moon in 1969, he described his surroundings as "magnificent desolation," an interpretation that has yet to be bested.[2] It's difficult to liken the Moon to anywhere familiar, because anywhere familiar is a place on Earth.

Even from orbit, Earth looks and feels like home. Astronauts report that staring down on our planet is one of the most exhilarating things about being in space. We belong here. Earth's razor-thin atmosphere, cloud tendrils, green-carpeted continents, and deep ocean blues beckon us. Not so for the Moon, according to Collins, who orbited it alone in his spacecraft but did not walk on it. There is no comfort to the "withered, sun-seared peach pit out my window," he wrote in his memoir, *Carrying the Fire*.[3] "Its invitation is monotonous and meant for geologists only."

Humans are sensory beings, and the Moon is a place devoid of any familiar sensory experience. If you were to visit, you might experience conflicting feelings of deprivation and overwhelm. Every time you went outside—in a spacesuit, of course—and every time you went back indoors and took off your gear, the Moon would bowl you over. You would feel lonely, hot, freezing, terrified, ecstatic, superhuman, and tiny, in a matter of moments or maybe all at once. Its topography, its innards, its atmosphere—everything about the Moon is different.

APOLLO 11 MOONWALKERS Neil Armstrong and Buzz Aldrin were the first human beings to experience selenic discomfort. Moon dust covered their spacesuits and boots, and it soon covered much of the inside of their *Eagle* lander, too. The pair were so annoyed by it that they slept in their helmets to avoid breathing in Moon all night. On later missions, astronauts noticed the dust scratched their sun visors and damaged the seals on the rock boxes they toted home. Moon dust caused a form of hay fever, making astronauts' eyes watery and itchy and their throats scratchy and sore. Unlike Earth dust, which is mostly made of organic material, Moon dust is all pulverized rock—and no water or wind exists to soften the dust grains' edges. It was like breathing in sandpaper.

But the astronauts were lucky that this was nothing more than a nuisance. NASA scientists had warned the astronauts that Moon dust might be reactive in oxygen. Aldrin and Armstrong were told to be cautious about their contingency sample, a small scoop of Moon that Armstrong tucked into his pocket moments after stepping out of the *Eagle*. After coming back inside, Aldrin and Armstrong watched the dust carefully as the *Eagle* cabin pressurized. If anything started to smolder, they were supposed to open the hatch and throw it out. But both men were completely coated in it.

"The stuff seemed to stick to things and stay there," Aldrin said.[4] "There was no hope of getting that off." If anything was going to ignite, it would be their suits.

The dust turned out not to be reactive in oxygen, but it did smell that way. The Moon has an acrid aroma, like fireworks that have just gone off. That is how Aldrin described the scent in the capsule after he and Armstrong came back inside from their brief sojourn and took off their helmets. Armstrong described it as "the smell of wet ashes," like a campsite at bedtime after you've doused the fire.[5] Apollo 17 astronaut Harrison "Jack" Schmitt has called it the smell of gunpowder.[6]

The Moon is constantly bombarded by sunlight and radiation from other stars and cosmic sources, and it's pummeled by asteroids in a process called "gardening." All this action tears apart atoms in the "regolith," the technical term for Moon dust. Lunar regolith is about 43 percent oxygen, so most of the atoms being shattered are oxygen atoms. The same is true of gunpowder. When it ignites, chemicals in the gunpowder release copious oxygen, further fueling the blast. What the astronauts smelled was the lingering aftermath of atoms being torn apart by tiny invisible bullets of radiation.

This is still a matter of scientific debate in part because the Moon rocks don't have a smell anymore. When a scientist opens a bag with a Moon rock today, no matter how carefully it was chipped and packed up for distribution by NASA's Lunar Sample Laboratory, there is no scent of the unknown. No one can say for sure why the smell fades once the rocks are exposed to humans, and to Earth.

ON THE MOON, after you got used to the smell of constant fireworks, you would notice the unceasing dryness. The Moon is a parched place, and you would dearly miss the omnipresence of water to which you have been accustomed your entire waking life. It would tease you every time you saw Earth. However familiar and beloved our continents and their mountains, Earth's land does not dominate the planet's features; from a distance, the water is what stands out, a blue beacon of serenity and warmth.

For most of human history, people believed that the Moon had oceans, too. Astronomers through the centuries believed the Moon's dark spots were actually lunar seas. Moon-fixated scientists in the seventeenth and eighteenth centuries believed this so fully that the list of features on its face are all named as oceans, lakes, and bays. The Sea of Tranquility, where Apollo 11 landed, was a real sea in the mind of Moon mapper Giovanni Battista Riccioli, the Jesuit priest who gave us the Moon's modern nomenclature in 1651. Collectively, the dark spots are called maria, from the Latin for "seas." In reality, as the Apollo Moon rocks taught us, the seas are vast plains of cooled lava.

While you would experience the Moon as a chalky, dry sea of emptiness, it does have water. Depending on what scientific instruments you believe, it has a whole lot. The trouble is that the water is locked up in the regolith as hydrated minerals, or may exist as ice that has been buried forever in craters that never see the light of day. Liquid water cannot exist on the Moon. With no atmosphere to keep water liquid, it would evaporate instantly, and its hydrogen would fly off into space. Any future Moon visitors hoping to access lunar water will have to be really talented chemists, skilled at liberating water from stone.

One reason so few places on Earth resemble the Moon is because of all the water on this planet. Earth's water softens and demolishes rock. Combined with wind, water is a destroyer of worlds. Entire mountain ranges rise and fall through the work of water. It also erases craters. Though the timing and duration of the beating are still up for debate, we know Earth was bombarded by asteroids long ago, and yet there are no battle scars to show for it. But the Moon, free of water and wind, preserves a record of its primordial pummeling. These craters play all sorts of tricks on the human mind.

In 2019, China landed a small robot called *Chang'e 4* on the Moon's far side, the first time such a feat had ever been achieved. A few months after the landing, a Chinese scientist named Long Xiao shared a video of *Chang'e 4*'s approach to the lunar surface. It was like watching an animation of a fractal. Each crater grew bigger and bigger in the lander's field of view, and then smaller craters resolved within the big craters, and those smaller craters also grew bigger until more tiny craters appeared within them, and on and on. In time, a visitor to the Moon would come to recognize the particular shapes and bowls of these craters, in the way I recognize my favorite mountains. But you would still have a hard time getting around because of them, and not just because the craters are tripping hazards. It would be hard to get around because your mind would have a hard time interpreting what your eyes saw.

The Moon's craters cast weird shadows, warping the landscape like a funhouse mirror. The extreme contrast between the darkness of space and the bright white of the Moon plays tricks on astronaut minds. During Apollo 12, Charles "Pete" Conrad and Alan Bean noted that the craters all seemed the same color, especially up close.

"The colors were deceptive. I can recall on the first EVA [moonwalk] looking at the materials around the LM [lunar module] and referring to them probably as gray-brown or gray-white," recalled Bean, who became a painter after leaving NASA in an effort to convey what he saw. "On the second EVA, in the very same places, although I wasn't really aware of it at the time, I referred to them as being light brown. I kept thinking that all the rocks had a light tan coating, whereas the first day I thought they had a light gray coating. My impression now is that the interior of all these rocks would be dark gray basalt, despite their very minor differences in texture, shape, etc. Also, both times we came into the LM our suits looked the same gray color. I saw only dark gray, never any of the browns that I'd seen outside."

The light fooled the astronauts in other ways. The Apollo missions landed when the Sun was low in the lunar sky, which helped the astronauts see crater shadows more starkly. Conrad and Bean landed near the "terminator," the line where lunar day turns to night. It was barely after sunrise, Moon time, and the Sun was only about five degrees above the horizon. Raise your fist, punch it toward the horizon—the Sun was about as high off the ground as your knuckles when the Apollo 12 astronauts set down.

Bean and Conrad touched down only 538 feet from a small lander called *Surveyor 3*, which had alighted on the Moon in 1968. This close landing was intentional, because scientists on Earth wanted to find out how the harsh lunar environment had treated the earlier spacecraft. The astronauts went to check it out, but they were worried the craters on the way were too treacherous.

"We may have a little trouble getting to Blocky Crater. I'm not sure whether it is an optical illusion or what, but the wall that the Surveyor is on looks a lot steeper than 14 [degrees]," Bean radioed Houston.

At one point the two moonwalkers tried to get a better sense of the dimensions. Conrad grabbed a grapefruit-sized rock and rolled it downhill. Later, commanders in Houston asked Conrad to climb into a crater (also confusingly called Surveyor) to collect some bedrock. "It's awfully steep," Conrad replied, declining the request. "I'll get you some bedrock from on the rim here."

Surveyor Crater, a 650-foot-wide impact basin, has a twenty-one-degree slope, a nice and easy downhill walk. The shadows had made the descent look more dangerous than it actually was.[7]

EVEN OUTSIDE THE steep craters, the Moon's surface undulates like the oceans for which its lava plains are named. In fact, it might be experiencing waves. Eons after it cooled into a solid orb, the Moon still undergoes some geologic activity.

Bean and Conrad left behind a seismographic measuring station, as did their successors on Apollo 14, 15, and 16. These instruments detect seismic activity deep within the Moon. Geologist Jack Schmitt, on his moonwalks during Apollo 17, noticed lunar geology that provided further evidence of seismic activity. On December 13, 1972, he and Gene Cernan parked their Moon buggy in a valley named Taurus-Littrow, within the Moon's Sea of Serenity. They were exploring a gray hill called the North Massif.

Schmitt noticed it first. "Hey, look at how that scarp goes up the side there," he said. "There's a distinct change in texture."

"Okay. Oh, man; yeah, I can see what you're talking about now. It looks like the scarp overlays the North Massif, doesn't it?" Cernan said.

"Yeah," Schmitt said. To Houston, he narrated the sight: "The appearance you have of the scarp–North Massif contact is one of the scarp being smoother-textured, less cratered, and certainly less lineated. And I wouldn't be a bit surprised if it's, as Gene says, younger."[8]

He meant that the scarp had formed after the mountain. Something had shifted the Moon's surface. He was right. In a study in 2019, scientists connected shake data from the Apollo seismic stations with updated lunar landscape imagery for the first time, and showed the Moon is geologically active today.[9] The Moon calves boulders. It forms rock piles. It experiences landslides that form escarpments like Lee-Lincoln, which appear like a shrugging shoulder. The Moon has fault lines, which experience regular moonquakes that are energetic enough to rattle an astronaut—or a future habitat.

The Moon has no mobile crust, unlike Earth. Its internal tremors are the result of tidal action between the Moon and Earth, and a vestige of its primordial heat. As the Moon cools—still, today, four and a half billion years after its formation—it contracts. Its crust wrinkles and collapses like a grape turning into a raisin.

These quakes appear to be fairly common. The four Apollo seismic stations counted twelve thousand seismic events, including twenty-eight shallow quakes, between 1969 and 1977.[10] The shallower quakes are similar to the types of tremors we feel on Earth. Spread out over eight years, that's still a quake every few days. Every few Earth days, that is. A day on the Moon is quite a different thing.

A "DAY" ON the Moon, meaning one full rotation and one revolution around Earth, takes twenty-seven Earth days, seven hours, forty-three minutes, and twelve seconds. We call this a sidereal month, for the time it takes the Moon to orbit once around Earth and return to the same spot relative to the stars. But because the Earth-Moon system is rotating around the Sun, it actually takes a little longer for the Sun to return to the same place in the Moon's sky. The synodic month corresponds to one complete cycle of phases visible from Earth. From the point of view of the Moon, the synodic month marks the time between successive sunrises* in the same spot on the Moon. No matter where you stand, this takes twenty-nine and a half Earth days.

Put another way, if you were standing on the Moon, it would take a full Earth month for the Sun to rise, set, and rise again. This also means daylight lasts for two weeks—and so does the night. You would need special equipment to survive this. Even some of the most sophisticated spacecraft we can build succumb to the frigid darkness of the lunar night, when the temperature drops to 300 degrees below zero Fahrenheit. During the daytime, the Moon is a scorcher: The average daytime high temperature at the equator is 246 degrees Fahrenheit.[11] Within some deep craters, a few of which harbor the ice you might need for survival, the Sun never shines at all.

During Apollo 11, Armstrong and Aldrin had a hard time sleeping in

---

* As you may have guessed, the English word "month" comes from the Moon. It derives from the Old English word *mōnath*, which has Germanic origin, and has many cognates in Dutch (*maand*), German (*Monat*), Latin (*mensis*), even Greek (μήν or *mḗn*).

their lander. The dust was annoying, but when they donned their Moon suits to escape it, they shivered with cold. The air-conditioning system in the suits was meant to keep them comfortable during the hot lunar day—inside the lander, it left them frigid. Anyone who ever visits again will need a life-support system that allows them to survive the extreme temperatures on the Moon.

The good news is you wouldn't mind walking around in what is essentially a wearable house. The Moon's gravity is one-sixth of Earth's, which effectively means everything feels lighter. You would weigh just 16.6 percent of what you weigh on Earth, so a spacesuit wouldn't be a burden. You might still have a hard time standing up, however; many of the Apollo astronauts fell flat on their faces after stepping onto the Moon's surface.

Modern studies show why this happened, and that it might be more than a trick of the light that fooled Bean and Conrad. In a 2014 study in Toronto, Canada, volunteers spun around on a giant rotating arm to simulate different gravity forces.[12] As they whirled and tried not to vomit, the volunteers were shown the letter p. They apparently read the letter as a p or a d, depending on which way they thought was up. They weren't tilted in their centrifuge; it was the gravity change that confused them. It turns out humans need to feel about 15 percent of Earth's gravitational force to sense which way is up. The Moon's gravity is just a smidge higher than this, at about 16.6 percent of Earth's. The low gravity and resulting disorientation might explain why it's so hard to walk on the Moon.

To make matters worse, time seems to stop up there. It proceeds according to the rhythm of your heart, and maybe the beeping of your spacesuit's life-support system, but if you could just stand there for an hour or two in silence, you would notice nothing about the passage of time. There are no long shadows cast by the noonday Sun. There are no changes in the angle of the light or the speed of the wind. Although you wouldn't consciously notice time passing, your body would feel it.

Typically, you don't notice time the way you notice scent or touch, but time perception is a sense all the same, and it pulsates through every cell in your body, and every cell in everything else that lives. Every form of life that we know displays some kind of time-dependent activity. This circadian clock is wound by the dependable cycle between light and dark—with some special lunar exceptions—because that has been the length of Earth's day for about as long as there has been multicellular life. When the light lasts for two weeks without end, the circadian clock becomes frazzled, at best.

Some elements of time on the Moon are more familiar. The Moon experiences solstices and equinoxes just like Earth does, but only at the Moon's poles is there anything resembling a season. The Moon is tilted on its axis by only 1.5 degrees, compared with Earth's 23.4 degrees, which is what gives us our four seasons. Temperatures do rise and fall seasonally at the Moon's north and south poles, where the angle of the Sun is always extreme. Though it is tilted only very slightly on its spin axis, the Moon is inclined relative to the Sun and the plane where the planets are found. When the Sun is above the Moon's equator, it's summer in the Moon's northern hemisphere, just like it is on Earth. When the Sun is below the Moon's equator, the northern hemisphere is in winter. The Moon also experiences the solstice, when the Sun appears to stand still and reverse its direction. Civilizations around Earth learned to measure these solar events and use them to mark time. Maybe future Moon settlers will erect solstice calendars, or build Earth phase calendars like some Neolithic humans once built Moon phase calendars to use the Moon as a guide.

In the lunar southern hemisphere "winter," more of the Moon's south pole is frigid enough to contain ice, which can be composed of water or even carbon dioxide (what we know as dry ice). As the Moon approaches its own equinox, some of that water is released into orbit.

THOUGH THE MOON has a version of a water cycle, there is no softly falling rain or snow. The sound of a thunderstorm might be one of the Earthly experiences you'd miss the most. There is no sound at all. The Moon is as quiet as quiet gets. You would hear yourself in your spacesuit, and yourself alone. Even if you tried making a racket, like clanging a piece of aluminum against your spacecraft, you wouldn't hear a thing. The Moon has no atmosphere to speak of, only an ephemeral "exosphere" consisting of some charged molecules and hovering dust. It is too thin for sound to carry.

Every Apollo mission both landed and launched its moonwalkers using rockets on the lander. The act of arriving on and blasting off the Moon kicked up enormous clouds of dust. Bean, on Apollo 12, is one of the reasons we know this. When he approached the earlier lander known as *Surveyor,* he noticed it had turned brown in its two years on the lunar surface. This happened because of radiation from the Sun and cosmic sources. But the Apollo 12 lunar module, named *Intrepid,* had sandblasted the *Surveyor* when it touched down just a few hundred feet away. Some of the brown was scoured away, like someone had scrubbed *Surveyor* with steel wool. Phil Metzger, a planetary physicist at the University of Central Florida, found that each landing blew lunar soil around, accelerating it between 400 meters per second and 3 kilometers per second.[13] That last number is important: 2.4 kilometers per second is lunar escape velocity. That's how fast something has to move to escape the gravity of the Moon and fly away. That means every Apollo mission kicked up enough dust, and pushed it around quickly enough, to send it into orbit. Some of that dust is still circling the Moon. Some of it is circling the Sun. And some may have even sprinkled back onto Earth, from whence it came, four and a half billion years ago.

THE LAST AND maybe most pervasive sensation you would notice on the Moon is an ethereal one. How many times in your life have you felt a sort of sixth sense? It is indefinable but unmistakable; you just know when the car in the lane next to you is about to merge, you can feel the presence of an animal behind you or a bird above you, you sense when you are not the only person in a quiet library.

This feeling is not one you would experience on the Moon. The feeling, instead, is a profound awareness that there is nothing, and there is no one. Everyone who has ever existed is up above, on Earth. Every being that ever lived and died and breathed and loved is distant from the Moon, instead sailing above you, appearing to go around you just like the Sun and the stars. Collins, on Apollo 11, was the first to experience this displacement. As he sailed around the far side of the Moon, out of touch with Aldrin, Armstrong, and Earth, he experienced a deep sense of solitude. "I am alone now, truly alone, and absolutely isolated from any known life. I am it. If a count were taken, the score would be three billion plus two over on the other side of the Moon, and one plus God only knows what on this side," he wrote. "I feel this powerfully—not as fear or loneliness—but as awareness, anticipation, satisfaction, confidence, almost exultation. I like the feeling."[14]

The Moon does not have humanlike feelings, but if a world can be described as lonely, the Moon would be. It is certainly empty, destined to face forever the world it came from and the world that, maybe because of the Moon, was blessed with air and water and life. The Moon will accompany us forever, but it will be lonely forever, if we treat it right.

CHAPTER TWO

The Creation

May the gods who dwell in heaven and the netherworld constantly
praise the temple of Sin, the father, their creator.

—The Nabonidus cylinder of Abu Habba (Sippar), col. ii.26–43a

IN THE BEGINNING, all was chaos. Before the skies were named, before
Earth existed, there was nothing but water, swirling around in a void
without form. But then something happened. The waters divided. The
roiling mix separated into fresh water, embodied by the short-tempered
god Apsu, and salty water, embodied by the goddess Tiamat. After their
sacred marriage, Tiamat gave birth to all the other gods of creation.

The younger gods were loud and obnoxious, and they kept Apsu
awake, so he decided to destroy them. Tiamat would have none of this
and alerted her older son, Enki, the god of wisdom. So Enki killed Apsu,
and built a home from his remains. In a great battle that followed, Tiamat herself was torn in half, and one half became the heavens, while the
other part became Earth.[1]

This story comes from the Sumerian Seven Tablets of Creation, one of
humanity's oldest origin stories, fragments of which are found on tablets from Ur, an ancient city whose ruins are located in modern Iraq. But
it shares many parallels with the utterly violent, riotous mess that was
the birth of the solar system and the formation of the Moon and its
Earth. The scientific version goes like this:

In the beginning, about 4.6 billion years ago, all was chaos within a
cloud of gas left over from a previous generation of stars. There was
nothing but molecules of dust and gas, swirling around in the void. The

star stuff drew closer together, and then something happened. The material started to collapse under its own gravity. The Sun ignited. *Let there be light,* a later creation tale says. Winds howled outward from the infant Sun, much more powerfully than the charged particles that stream through the solar wind now, and the gales pushed the remaining dust and gas around. The roiling mix eventually separated into clumps, which grew into larger piles, and eventually became the planets.

There were more than the eight we have today. Some of the Sun's original planets were probably consigned to oblivion. Gravitational interactions caused planets and planetesimals (planet embryos, basically) to knock into one another like billiard balls, and some likely exited the solar system, doomed to silently sail among the stars. We'll never know how many suffered this fate. But we do know that one of these primeval planets was completely obliterated. One of them, born in the same band of star stuff as Mercury, Venus, Earth, and Mars, is a world no more. It was probably about the size of Mars, roughly 45 percent the mass of today's Earth. Its name was Theia, after the Greek goddess who was the mother of Selene, the Moon. Theia was destroyed like Apsu and Tiamat. And Earth and the Moon made their home in its remains.

UNTIL APOLLO ASTRONAUTS landed on the Moon, dotted it with scientific instruments, and brought bits of it back home, we didn't know about Theia. We had no Earthly idea how the Moon got here, just a series of educated guesses. The Apollo missions totally rewrote the story of the Moon's origins. At the same time, Earth scientists began rewriting the story of Earth's own formation and singular geologic history, and they began to realize that the Moon has much to tell about Earth, too. The Moon's story is the shared history of our home planet, after all. And the Apollo rocks are still providing new clues. The Moon visits provided so much new material, and so many unexpected questions, that they have forced scientists to completely reimagine the making of the solar system more than once. Just as the Moon reflects Earth's light, its primary role in modern science is to tell our story back to us. The story is more than a scientific curiosity. The Moon's origins can shed light on how we all got here, and maybe even why.

Wouldn't it be wonderful to know? Why us, why here? Why not everywhere?

There are other rocky planets, but none are like Earth. Mars is also a slowly spinning terrestrial world, tilted on its side almost exactly as much as Earth is. But it lost its water and its atmosphere. And it has no Moon, just dinky captured asteroids. Venus is a faster-spinning rocky world with a dense atmosphere, but its cloak of clouds grew too thick over time, and choked the planet to death. If Venus ever had water, it's gone now. Mercury, too close to the Sun, somehow still harbors tiny amounts of water in the dark shadows of its deepest craters. But it is blasted by solar rays. Neither Venus nor Mercury has moons.

Why do we? What was it about Theia, the original Earth, and their mutual destruction that would give rise to this planet? Why did we end up with a huge Moon, one-fourth of Earth's own heft? What happened in that cataclysm that resulted in a paired system of worlds, one dry and completely dead, and one drenched in water and life?

THE MOON'S APPARENT size compared to the Sun, and the fact that we have solar eclipses, led many ancient humans to think the Moon was placed in the sky alongside the Sun for a reason. The Navajo people of the American Southwest believe, like so many cultures, that both were created at the same time, and for similar purposes: one to light the day, one to light the night. Many religions through the ages thought the same thing. People also came to assume the Moon was given to us as a timepiece. Plato even asserted that the succession of days and nights, lit by the Sun and Moon, taught us how to count—and how to think.

Over time, the notion that the Moon was purposefully and wonderfully made morphed into a scientific theory. Philosophers and scientists reasoned that the Moon was made along with Earth, forged from the same primordial matter that makes up the Sun and other planets. The Moon must have formed at the same time and in the same place as Earth, and stayed here owing to Earth and the Moon's mutual attraction. Scientists promoted some version of this until the turn of the twentieth century C.E., when a pioneering astronomer named George Howard Darwin came up with an alternative explanation.

George—son of the famous biologist Charles and his wife, Emma—began his theory by thinking about the tide. His dad had written that the Moon's control of the tides may have been responsible for the origin of life, bubbling up in "some warm tidal pool." But George wondered whether the tide had anything to do with the existence of the Moon itself. Thanks to the work of earlier scientists, he knew that Earth's rotation is slowing down, ever so slightly, through tidal interactions with the Moon. The slowdown of Earth's daily spin means Earth is losing angular momentum.[2] Angular momentum is always conserved, meaning it can only be gained or lost if something else gets involved. George Darwin understood that this conservation of angular momentum means that, as Earth slows, the Moon is moving away. If it's receding all the time, then it would have been a lot closer in past epochs.

Darwin ran some calculations and found that in the not too distant past, Earth and its Moon would have been practically touching, and Earth's day would have been four hours long. Somehow, the Moon was being flung away. In 1899, he proposed a tale of fission formation: The Moon calved like a glacier, breaking off from Earth into something apart. He reckoned the Moon was probably shorn from somewhere in the Pacific Ocean, which is why that ocean is so deep.[3]

This is not what happened, but George deserves credit for coming up with the basic outline of an idea now found to be true: The Moon was not formed separately, nor was it baked alongside Earth in some strange process of twin genesis. The Moon came *from* Earth and shares its history.

The factual basis for George Darwin's story comes from the Apollo missions. It comes from a few pebbles Neil Armstrong grabbed in his first moments on the Moon. It comes from later, more famous Apollo rocks. And it is still being unraveled through especially beautiful, oddly alien rocks, like the pearlescent, green-freckled specimen designated troctolite 76535.

Many Moon rocks are white and gray chunks of something called anorthosite, a strange and low-density type of material that forms when minerals crystallize within molten rock. They are among the first samples Neil Armstrong gathered, and because they were so unusual, Apollo missions kept going back for more. These glittering Moon bits are special in part because they are so pure. Their refinement, especially compared to average Earth rocks, surprised geologists in the 1970s. A typical rock on Earth comes in one of three flavors: igneous, which forms when molten rock cools; sedimentary, which is deposited by water and wind over eons; and metamorphic, which is the other two rocks transformed in the crucible of the ages. All of these rocks can contain multitudes of minerals. But Moon anorthosite does not. It's made almost entirely of a mineral called feldspar.

As rocks go, this stuff is a dime a dozen, both figuratively and literally. Feldspar is so common on Earth that it is frequently used in kitchen cleaning powder because it is crumbly and low density. On the Moon, these properties make it special. The anorthosite rocks coat the surface of the Moon because they floated there. They bobbed in a sea of melted Moon like an iceberg in Earth's oceans. Heavier stuff sank within the oozing magma and formed the Moon's core, while the anorthosite crystals floated to the top, forming a thin crust like the skin on a day-old pudding cup. As the freshly baked Moon cooled, the rocks were locked in place, only to be liberated later during asteroid strikes.

In order for this material to separate from the rest of the Moon, for anorthosite to make a white, Moon-ish pure crust, it needed an ocean to float in. It needed an entire Moon's worth of rock ocean. And a magma ocean spanning the entire Moon could only form through some incomprehensible violence: like a meeting with Theia.

THE STORY OF Theia is unique in the solar system, so far as we can tell. It begins 4.35 billion years ago, give or take a couple hundred million Earth years. And it doesn't begin with Earth, not really. It begins with Earth 1.0.

If you were able to visit this alien world, you would recognize nothing. Earth 1.0 spun like a dervish, rotating through one day and night every few hours. The infant planet's steaming rock was constantly buffeted by high winds. There was land, sort of, and maybe some water. The night sky would look different, with constellations in slightly altered arrangements, but you would notice the Milky Way stretching overhead, and Jupiter shining in the southern sky. And you would see a glittering orb above: A few days before the collision, Theia would have been the same size in primordial Earth's sky as the Moon is in ours.

The collision was inevitable, fixed by fate and gravity. Earth 1.0 orbited the Sun from a distance of about ninety-three million miles, roughly the neighborhood where it is now. Theia was close to the Sun, too, right in the so-called Goldilocks zone, where conditions are not too hot, not too cold, but just right for water to remain liquid. But this neighborhood was not wide enough for two rocky worlds.

Objects that formed around the infant Sun were zipping about much faster back then. As the moment of impact approached, Theia sped toward Earth 1.0 somewhere between the ludicrous speeds of 20,000 miles per hour and 8,900 miles per hour. At the slow end, that is about four times faster than a bullet fired from a .30-caliber hunting rifle. This analogy is imperfect for an entire planet, but a discussion of speed is at best an academic way to convey the scale of the calamity.

IF YOU CAN see outside right now, you can try to picture what would have happened when the world ended. Look around. Is there a skyscraper in your line of sight, or a neighbor's house, maybe a tree? Find the sky. Is it blue? Maybe you see a jet contrail, or even an airplane streaking by, its tail shining in the sun. Now imagine: Here comes Theia. First the airplane vanishes, then the contrail. The tree and the buildings burst into flames as an angry red mountain—no, an upside-down world—descends, closing off the heavens from east to west. The Sun still shines, so you can see fine detail in the upside-down world as it approaches. There are rivulets of lava and craggy peaks shadowing deep valleys.

Then the entire horizon dims to a livid red glow as Earth 1.0 begins to moan and tremble, shockwaves rattling through its crust and deep into its mantle. Theia's gravity pulls on Earth 1.0 and Earth 1.0's gravity pulls on Theia until the sky closes off for good, as the crust of each planet meets the other. Devastating seismic waves ripple down through the mantle of Earth 1.0 and Theia, and both planets splatter apart.

TWO YEARS AFTER the final Apollo missions, selenographers held a conference and finally published a comprehensive theory for this story. A Harvard professor named Reginald Daly had proposed a giant-impact theory in 1946,[4] but the idea didn't gain much traction until Apollo astronauts hauled down all that anorthosite to show that the Moon was, indeed, once liquid magma.

The Caltech planetologist Dave Stevenson was an undergraduate at Cornell University during the conference, in 1974, and the idea stuck in his mind. A decade later, he attended the Conference on the Origin of the Moon in Kona, Hawaii, and by then he and the majority of selenographers thought they had solved the mystery of the Moon's origins.

"Without colluding, many of us arrived at this meeting saying, 'Hey, this is the right story: a giant impact.' From a physics standpoint, it looked attractive. It's a trivial calculation," Stevenson told me. Two worlds smashed, utterly ruined each other, and their remnants eventually calmed and cooled into the two new worlds we have now. It made a lot of sense. But rocks are not simple things. The story of Theia and Earth 1.0 is complicated. Even as they built the theory, some geochemists were unsatisfied. From the earliest days of the giant-impact hypothesis, the story of the rocks and the narrative brought to life by physics have not quite matched.

IN THE BEGINNING, when all was chaos, rocky worlds were forming and colliding and vaporizing one another continually.

The basic outlines of this primeval pandemonium come down to us from the great German philosopher Immanuel Kant. In Kant's metaphysics, reason is the source of morality, and in his astrophysics, chaos is the source of creation. In 1754, the Royal Academy of Sciences at Berlin awarded him a prize for a treatise called *Examination of the Question Whether the Earth Has Undergone an Alteration of Its Axial Rotation,* in which he wondered whether any outside forces had ever acted upon the spinning Earth. He considered the tides, and argued that the movement of water would act to slow down Earth's rotation. Eventually, Kant reasoned, Earth's rotation would slow down so much that its spin would match the rate at which the Moon goes around it. This would result in Earth always showing the same face to the Moon—just as the Moon always shows the same face to us. He was right, though the mathematics underlying this idea would not be explained until George Darwin. A year after his Moon essay, Kant advanced a new theory of how not just the Moon but the entire solar system came to be, in his treatise titled *Uni-*

*versal Natural History and Theory of the Heavens, or an Essay on the Con-
stitution and Mechanical Origin of the Whole Universe, Treated According
to Newton's Principles.* The philosopher wrote at roughly the same time
as Pierre-Simon Laplace in France and William Herschel in England,
and all three posed slightly different variations on the same theme. But
Kant's version of our collective origin story is the closest to the truth.

He argued that the primitive chaos divided itself into isolated
masses—like fresh and salty water, like Apsu and Tiamat. He imagined a
great number of particles swimming around a center of gravity. "This
body at the middle point . . . is the sun, although at this time it does not
yet immediately have that flaming glow, which breaks out on its surface
when its development is fully complete," he wrote.[5]

KANT'S "NEBULAR HYPOTHESIS" is basically correct: The Sun and planets arose from a chaotic swirl of dust and gas that slowly collapsed and
grew through the forces of gravity.

Since Apollo, scientists have advanced a few versions of this story's
particulars. Today, many scientists think the solar system arose from a
sort of cosmic curdling in a process called the streaming instability. Gas
and dust drifted around the Sun like snow around a tree, concentrating
in certain areas—controlled by things like pressure and temperature—
and then collapsed and condensed into small, compact objects. These
dust bunnies eventually grew into protoplanets a few kilometers across,
which amassed more material over time. Some scientists now think
that after the first dust bunnies formed, they quickly accumulated more
pebbles and dust that swirled around them and very rapidly mush-
roomed into the planets we now know.

This was a violent time. The dust bunnies and other planetary crumbs
collided, melted, recombined, and gradually formed larger balls. Eventually these grew big enough to clear their neighborhoods, meaning
they consumed all nearby crumbs, or their gravity swept the crumbs far
away, toward the eventual realm of asteroids and comets. As the infant
planets got bigger, their insides heated up, which allowed the metals in
their rocks to flow and differentiate into lighter and heavier elements.

Some of these infant worlds stopped growing after a short while, and
became the rocky planets Earth, Mars, Mercury, and Venus. Others con-

tinued hoovering material until they metamorphosed into the humongous gas giants Jupiter, Saturn, Uranus, and Neptune. Larger crumb piles orbiting those planets grew into the gas giants' suites of moons.

A great cosmic sorting thus took place. The material that made the planets drifted and settled according to its location around the Sun. This location is reflected in the planets' chemistry.

Everything in the visible universe is made from atoms, which contain a nucleus of neutrons and protons and an outer shell of electrons. But atoms of the same element can come in different sizes. Sometimes an atom can tack on a spare neutron, making it heavier, in a manner of speaking. It is still the same element, because it has the same number of protons, but it has a greater mass. An atom with a spare or a scant neutron is called an isotope. Scientists can count the number of isotope variants in an object and find out about its nature based on those numbers. An object with a higher proportion of heavy atoms is more radioactive. For the people who study chemical properties, an object's ratio of heavy to light isotopes can reveal information about its nature, and even how it was formed. In planetary science, an isotope is a bit like a person's accent, giving away its origins.

Closer to the Sun, where it's warmer, the molecules of lighter elements were more likely to heat up and escape during the formation of the solar system. Farther away, where things are colder, rocks were able to keep more of their water and other lighter substances. This simple chemistry partially explains why the planets are all so different. Mars is so chemically distinct from Earth that you can identify Martian meteorites just by looking at their oxygen isotopes, for example.

When Apollo astronauts brought the Moon rocks home, we discovered that the Moon was also different from Earth. Those pure white anorthosite crystals were distinct, for one thing. And Moon rocks seemed to have different numbers of oxygen, titanium, and other elements' isotopes. Based on these chemical accents, scientists who favored the giant-impact theory believed the Moon was made from the remains of the impactor. It was Theia reborn.

It was a nice theory. According to these scientists, the Moon came from the poor obliterated world that collided with Earth 1.0 early in the solar system's history. The name Theia, mother of the Moon, was a perfect fit. The notion came to be called the giant-impact hypothesis. It explains many peculiarities of the Earth-Moon system, from its angular momentum to the sizes of the two worlds.

But modern scientific techniques would begin to cast doubt on the story's simple outlines.

In 2001, Swiss researchers remeasured the Moon rock called troctolite 76535 and thirty other lunar samples using sophisticated equipment that sifts through isotopes.[6] They found that the rocks' oxygen isotopes were indistinguishable from those of Earth rocks. In the years since, geochemists around the world have taken a crack at a whole host of other elements and obscure metals. Titanium, chromium, rubidium, potassium, tungsten, and other materials all look pretty much the same, whether the rocks are from Earth or the Moon. When scientists measure these elements, they usually look for isotopes that have extra neutrons, the ones that are slightly radioactive. These isotopes can be used as a clock. This is the principle behind radiocarbon dating, which archaeologists commonly use to find out how old something was when it died.*

During reexamination of Apollo samples, isotopes of tungsten looked particularly strange. Tungsten-182 is a daughter[†] of an element called hafnium-182. That means tungsten-182 acts like a clock, just like carbon-14, albeit for much, much older things; this clock allows scientists to determine when a planet formed. But the ratios of tungsten and hafnium are essentially the same in Moon rocks and Earth rocks. Very unlikely coincidences would be required for the rocks that make up our planet and our Moon to have formed from the very same stuff at the very same time.

---

* Scientists study an isotope called carbon-14, which has two extra neutrons in its nucleus. All living things absorb normal carbon-12 and unstable carbon-14, and this absorption stops when the organism dies. The carbon-14 then decays, with a so-called half-life of 5,700 years, but the carbon-12 remains constant. The less C-14 something has, the older it must be.

† Radioactive decay occurs when an unstable isotope, also called a nuclide, emits spare electrons and changes into a different element. The original isotope is called a parent, and the version that forms during radioactive decay is called the daughter. The daughter might be stable, or it might also decay, but on a different timescale from the parent.

From the point of view of Theia, this does not make much sense. If Mars is so obviously different from Earth, if it has a distinct chemical accent, then Theia should, too. Planets contain the fingerprints of their birthplace, and Theia was born far from Earth. A Theia with an accent identical to Earth's would be a monumental coincidence. But Theia is a ghost, so we cannot interrogate its rocks.

GIVEN ALL THESE chemical similarities, some scientists began to wonder if Theia never existed. Some wondered whether the early Earth's environment might have been more crowded. All other mooned planets have more than one, after all, and some even have rings. Why should we assume Earth has always had just the one satellite? One theory suggests that Earth's Moon is not the original Moon, but a compendium of at least a dozen individual worlds. During Earth's youth, while the incandescent planet cooled amid a hellscape of rocks and Promethean fire, errant rocks smaller than Theia could have shorn away bits of the planet. These projectiles would have knocked off enough rock to form tiny wisps of Earth rings, which would combine over time, something we know occurs inside the rings of Saturn. Those ring-forged moonlets would eventually coalesce, forming the Moon we know today.

Though the theory is compelling, simulations still can't explain how the many moonlets could have combined into such a huge secondary world. What's more, the physics that describes Earth, the Moon, their rotation speeds, their distances, and many more characteristics can only be explained by another rocky world smashing into Earth 1.0. Physics insists that Theia must have been real.

And yet a Moon and an Earth with essentially identical chemistry must have formed from the same material, which no one can explain. So scientists have been trying to reconcile the messages from the Moon rocks with the requirements of physics—the need for an impactor of a certain size, traveling at a specific speed. Sarah Stewart, a planetary scientist at the University of California, Davis, is leading the charge. She and her colleagues have come up with several ways to imagine, and then simulate in computer code, the most horrendous and most consequential day in the history of this planet: the day Theia came to Earth.

Stewart is a MacArthur "Genius" grantee, a computer modeling expert focused on colliding worlds, and the type of magnetic personality who enraptures people with discussions of seemingly boring topics—like isotopes. When I met her at a conference of lunar and planetary scientists, she was wearing a chalk-white pantsuit and matching white cape, adorned with a glittering enamel brooch depicting the full Moon. Her lab is equipped with custom-built cannons that smash small rocks together at ludicrous speeds, simulating the dreadful conditions that result from clashes of planets and asteroids. Her computer simulations can mimic the world-rending collision of Theia and Earth 1.0, and the clouds of crumbs that eventually became planetesimals and planets.

In 2012, Stewart and a colleague named Matija Ćuk proposed a new story of the Moon's formation.[7] Their computer simulations suggested that when Theia collided with Earth 1.0, our natal planet rotated through one day every two to three hours. The violent, head-on smashup sheared away a huge portion of Earth, obliterated most of Theia, and mixed the remnants sufficiently enough to build an Earth and a Moon from basically the same ingredients. Their cores would be different, with the Moon incorporating Theia's innards and Earth keeping its own. But the rocks that we can study—the rocks that floated to their surfaces and were excavated by other asteroids—those rocks would look identical. This version had problems, too, however. Earth is not a whirling dervish now. As Immanuel Kant and George Darwin knew, it is slowing down thanks to interactions with the Moon. The Apollo astronauts left scientific instruments on the Moon that let us record the rate at which this slowdown is occurring. According to the rate of change, not enough time has passed to slow Earth from a three-hour day to our current twenty-four-hour rhythm. Something else would have had to transfer some of the angular momentum away from Earth. There are calculations involving the Sun that can account for this change, but it's not really enough, so the story of the Moon's formation was not quite complete. Competing computer models churned out variations on this theme for several years, and it was into this melee that Stewart and her student Simon Lock waded in 2018, updated code in hand.

"Everyone was still stuck in this idea that you form this planet and this disk. Maybe it was a different impact, you produce more vapor, but the fundamental structure is the same. What we found is that's not true," says Lock, who is now a planetary scientist at the University of Bristol in England. Something else happened instead.

The day Theia came down to Earth 1.0, Theia did not just shear away part of our world. Both worlds were completely torn apart. The devastation was complete, and in its aftermath, there was no ring. There were no naked planet cores floating in space. There was no planet and no moon. Instead, both Earth 1.0 and Theia were blasted apart into a super-heated cloud of dust. Their vaporized remains swirled into a fast-spinning, bagel-shaped bulging disk, a short-lived structure previously untheorized in planetary science. The Promethean hellscape of this structure defies our previous understanding. The cloud spun so quickly that its outer edge reached a point called the corotation limit, which essentially means it went into orbit. The thing is too big and diffuse to rotate like a normal planet; instead, at the outer edge of the cloud, the vaporized rock spun so fast that it took on a new structure, with the disk circling a hot inner region. But the disk is not separated from the central region like Saturn's rings, or like anything else any scientist had ever imagined. Every region of the cloud formed molten-rock raindrops, which Stewart and Lock initially called a continuous mantle-atmosphere-disk structure—a MAD structure. Earth 2.0 and the Moon cooled and coalesced in this cloud, like eggs poached in a pot of boiling water. The seed of the eventual Moon would have formed within just a year, and the two bodies would have remained in hell-cloud form for just a century before settling into the paired worlds we recognize today, according to Lock.

Stewart and Lock decided this unusual planetary nest needed a new name, so they dubbed it a synestia. It's derived from the Greek goddess Hestia, champion of the hearth and home, and the Greek prefix *syn-*, which means "together." Our home, together.

The particulars are still debated in planetary science conferences, but most scientists agree that the Moon probably formed fast. In a 2022 supercomputer simulation, Theia and Earth 1.0 collided at twenty thousand miles per hour, and the wrecked planets launched a piece of themselves into orbit. The Moon formed within mere hours of the impact, according to this theory.[8]

To understand the fine details of how the Moon and Earth came to be, scientists will need fresh samples from the Moon. Meanwhile, theorists will keep trying to understand the Hadean conditions from which our Earth and our Moon were born. Their work may describe more than the birth story of the Moon.

We still haven't nailed down the particulars of Kant's solar system origin story. None of the current planetary formation and accretion theories truly capture the solar system we see. This is true for planets in other solar systems, too. No single idea can explain the huge variety of worlds we know exist, nor the variety of places they are found.

Planets seem to coalesce very quickly, both in our solar system and around distant stars. This suggests they don't often form through a core-accretion process, because many eons are required to mash crumbs together into something the size of a world. Jupiter, for instance, contains the vast majority of the material left over from the birth of the Sun. The planet has a huge core, according to measurements from the *Juno* spacecraft, which began orbiting the gas giant in 2016. Building Jupiter's huge core would take a long time—too long, according to the best simulations of this process. The material around the infant Sun would have disappeared in less time than would be required to bake such a humongous world. What's more, Jupiter is in a weird location. In most other star systems, giant planets like Jupiter form in the outer reaches, and might later migrate inward. But Jupiter, which is more than twice as massive as the rest of the planets combined, is somehow mixed in among all of them and is the closest to the Sun of any gas giant.

Other models also fail to explain the panoply of planets we know are out there. The curdling mechanism of streaming instability would be able to produce worlds only in certain locations around a star. But they are found here, there, and everywhere. Some planets orbit their hosts in a couple of Earth days, and other star systems are home to far-flung, frigid worlds that make Jupiter's orbit seem snug. None of our theories quite add up to the universe we can see through our telescopes, both nearby and unimaginably far away.

There are some observational and mathematical reasons for this lack of understanding. Telescopes just can't make out objects that fall in the spectrum between specks of dust and protoplanets the size of the Moon. To understand why, we have to wade into the laws of optics.

EVERYTHING YOU SEE represents just a tiny sliver of the electromagnetic spectrum. Most of the universe is hidden from view, even the normal stuff, not just the mysterious dark matter no one yet understands. The right telescopes are technologically capable of seeing most of the light we cannot, but Earth gets in the way. Its atmosphere absorbs infrared, ultraviolet, X-ray, and gamma-ray radiation, which is part of the reason we can live here without being fried or having our DNA ripped apart. The atmosphere also blocks some visible light, which is why the Hubble Space Telescope, the James Webb Space Telescope, and other observatories are orbiting above us. In very high and dry places, like the mountaintops of Hawaiian volcanoes and the Andes Mountains in Chile, large radio telescopes can pick up long wavelengths of the electromagnetic spectrum. When I visited the Atacama Large Millimeter/submillimeter Array (ALMA), medics gave me an oxygen tank, which the engineers wore in backpacks. The lack of atmosphere so helpful to astronomy can wreak havoc on your brain and body.

Radio telescopes can sense information from almost any source,* from the Sun to cold, dark objects that emit no visible light. The signals that let you make a call on your smartphone are a billion billion times more powerful than these faint cosmic emanations.

Because radio telescopes detect long wavelengths of the electromagnetic spectrum, their light-collecting surfaces have to be quite large. Modern radio telescopes cheat this law by combining several different dish antennas and synchronizing their data using a supercomputer. ALMA, perched at sixteen thousand feet above sea level in Chile's vast Atacama Desert, uses fifty-four such dishes to "see" objects from millimeter-sized and centimeter-sized dust, up to small, cold pebbles.

But even with ALMA's keen vision, there is no way to make out a kilometer-sized embryonic world, at least not yet. To see something a kilometer across, you would need a kilometer-sized radio telescope array. There is one under construction in South Africa, but even the Square Kilometre Array may not be able to see a planet in the process of growing from a crumb to a world. There may be some stages of the pro-

---

* This is why many astronomers favor building a radio telescope array on the Moon's far side, where it would be uncontaminated by the deafening roar from Earth.

cess we will never observe. The upshot is that nobody really knows what the middle phase of a planet's development looks like.

Mathematically, computer simulations have a hard time separating the motions of gas from the motions of dust inside baby star systems. Most simulations assume the material all moves together in lockstep, but maybe it doesn't. Maybe gas swirls around the dust grains, pushing the dust together the way water shoves flood debris into eddies and pools. "Most models haven't actually calculated this collisional mixing," said Sarah Stewart.

Enter the synestia.

Maybe, warm clouds that form in titanic collisions like the one that forged Earth 2.0 and the Moon are responsible for this missing link in planetary formation. As hot vapor in a synestia cools and contracts, it provides an inward flow of collapsing gas. The dense clouds of planet crumbs are protected from stellar winds long enough to form larger clumps, and eventually the parent bodies of planets.

"They act like a security blanket, or a shield to protect them from turbulence in the surrounding nebula," Stewart said. This cosmic cuddling could also help explain why objects that form together look so similar.

It's possible that a synestia is just a phase of a planet's life, like its birth and its inevitable destruction. Maybe all of the planets started out as a synestia. Every planet that is violently whacked by another world forms a synestia, according to Stewart, and we know there was a lot of knocking around long ago. These hellish structures might be common, especially in the early years of a star's natal cloud.

If this is true, the story of the Moon's formation is much more than the story of our silvery sister. The origin of the Moon is the story of creation writ large. Maybe the Moon's birth story can tell us about the genesis of not just Earth, but all possible worlds. We don't know yet. But the Moon still guides these questions, for this planet and all others.

$$\circ \; ^\bullet$$

WE KNOW THAT the obliteration of Theia led to the Moon and Earth, but minor bombardments by space rocks were common and lasted for eons after Theia's remnants were remade into Earth and the Moon. Mercury, Venus, the Moon, and Mars were all pounded by uncountable asteroids for ages. Just look at their cratered, dimpled faces, which have remained unchanged for billions of years and still show the scars of this primeval pummeling. Only Earth has no evidence of this cataclysmic history. That is not because Earth escaped the drubbing. It is because Earth's crust is alive with the action of plate tectonics.

Look at a globe, or even a flat map of the world. The puzzle-piece shapes of Earth's continents, especially South America and Africa, are immediately obvious. Just slide South America over to the east, and Brazil's right shoulder nestles right under Africa's left. Brazil's beaches neatly meet Ivory Coast, Ghana, Togo, Benin, and Nigeria.

Explorers dating to the sixteenth century noticed these shapes, and suggested that the continents were originally whole and later somehow torn apart. The earliest example of this notion probably comes from Abraham Ortelius, the Flemish mapmaker who published the first modern atlas in 1570. Many other geologists and explorers repeated this theory in the intervening centuries, even proposing (accurately) supercontinents that broke up sometime in the past. But no one could explain how the continents separated.

To some scientists, the idea finally started making more sense after George Darwin proposed lunar fission. Based on his theory, many scientists in the late nineteenth and early twentieth centuries speculated that Earth's features reorganized after the calamitous calving event, with the puzzle pieces of Earth's continents a direct result of the Moon's departure. Around the turn of the twentieth century, people put forth several scenarios. Geophysicist Osmond Fisher argued that the deep Pacific Ocean was a scar left by the Moon's absence. Astronomer William Pickering argued in 1907 that when the Moon flew away, Earth's crust slumped into the depression it left behind; this tore Africa from South America and ripped open the Atlantic Ocean. Geologist Frank Taylor thought the Moon was captured, not calved, and that the continents shifted over time because of the tidal interactions that resulted from the gravitational relationship between the Moon and Earth.

The man who finally explained this continental puzzle also tried to understand the face of the Moon.

IN THE SPRING of 1916, Alfred Wegener, a German astronomer and atmospheric physicist, was thirty-five and home on two weeks' leave from
service as an infantry captain on the western front of World War I. On
the afternoon of April 3, 1916, a fireball streaked across the skies north-
west of Marburg, where, in peacetime, Wegener taught meteorology at
the university. Captivated, he spent his two weeks' R & R traipsing
through west-central Germany and interviewing witnesses. Alfred and
his wife, Else, knocked on doors in the small hamlet of Treysa, which he
determined was the closest town to where the space rock would have
landed. Eventually, in January 1917, the meteorite was recovered, only a
few hundred yards from where Wegener calculated it would land.

Wegener's interest in the meteorite fueled his interest in the Moon's
own apparent craters. By Wegener's time, people were still not sure how
the Moon's surface came to be so dimpled. Scientists debated several
ideas, like a so-called bubble hypothesis in which the Moon's surface
erupted like a multitude of boils; volcanism, a theory favored by Immanuel Kant and the pioneering naturalist Alexander von Humboldt, which
assumed the craters were ancient caldera; and even ice ages, in which
the craters were all ancient glacial moraines. Many people favored a
Darwinian tidal model, in which the Moon's interior was rent by violent
tidal pressures* in the days before it became locked with one face to
Earth. With each rising tide, cracks in the Moon would leak molten material, which would congeal and form a solid circular ridge.

All of these were wrong. And in the fall of 1918, now home from the
war, Wegener set out to prove it.

As historian Mott Greene describes him, Wegener knew more about
selenography than most geologists, and knew more about geology than
any selenographer.[9] Like Johannes Kepler, he stood above the field of
research that he'd carefully surveyed, and incorporated other scientists'
findings into his own new, unorthodox theories. Wegener owned a copy
of Kepler's fictional lunar opus, called *Somnium,* and carefully annotated the sections where his fellow German astronomer had described

---

* Tidal pressure is responsible for volcanism on Jupiter's moon Io, but this wasn't known
until 1979, when astronomer Linda Morabito studied *Voyager* flyby images of Io and spotted a volcanic plume streaming into space.

craters as the result of impacts. In a series of experiments performed in 1919 and 1920, he would pile cement powder on a table, sprinkle it with water, and let it set into concrete. Then he would dump more loose cement powder on top, and toss a tablespoon of additional cement powder on that. The result was a tiny crater. Again and again, he made and photographed cement-powder craters of different sizes, noticing that some formed central peaks and some did not, depending on the amount of impactor material.* He compared photographs of his mini craters to real Moon photographs, which appeared increasingly detailed through the most recent telescopes. And he realized he was onto something. "Typical lunar craters are best explained as impact craters," he wrote in a 1921 report.[10]

Wegener was correct about lunar craters, and was mostly correct about the forces of plate tectonics. Rejecting all of his predecessors' ideas about Earth's movable face, he began circulating a hypothesis that Earth's continents drift around on their own accord. He argued that Earth's crust floats on top of its interior—just like Neil Armstrong's anorthosite rocks originally floated on a hot Moon, though Wegener didn't know about that. Wegener suggested that this happened through tidal forces resulting from the gravitational pull of the Moon. He also proposed that Earth's equatorial bulge forced the continents to drift away from the poles. His nemesis, British geophysicist Harold Jeffreys, demonstrated that these forces were inadequate, and Wegener's ideas languished until sonar developed for submarines during World War II helped scientists discover the phenomenon of seafloor spreading. This is the mechanism that actually explains continental drift—though good luck finding a geophysicist who can tell you simply why that happens, or

---

* You can try this at home. Clean off a kitchen counter and take out the bag of all-purpose flour you probably have tucked in the pantry. Dump out about a cup, covering an area about six square inches. Using a teaspoon, scoop a little more flour out of the bag, packing it tightly into the spoon. Then, holding your arm about six inches above the floured counter, flip over the teaspoon, dropping a floury meteorite onto your pretend regolith. Watch what happens. Is there a central peak inside your new crater? How tall is the crater ring, and how wide? How far do the wisps of flour radiate from the point of impact? Try changing the crater's characteristics by making a thicker surface, or by adding more or less flour to your meteorite. It's okay if your kitchen gets messy. It's science.

why there are plates in the first place. As Earth's crustal plates converge and slide under one another, magma rises between the fractures in their meeting grounds (mid-ocean ridges) and forms fresh seafloor.

Wegener would not live to see his theory of plate tectonics accepted by the world's scientists. But his theory of lunar craters met with a much more receptive audience. After Wegener, most scientists believed, rightly, that the Moon's surface looks like a pizza because it has been walloped by asteroids, then and now and forever.

THE MOTHER OF all collisions, the Theia impact, changed the course of our planet's history. The collision remixed the very materials that make up Earth. Theia may be a ghost, but it is not gone.

Theia may have added a veneer to Earth's mantle, endowing it with an extra load of elements, like gold, palladium, and platinum, and other materials. These would typically bond to iron and sink to the core, but they are found throughout Earth's mantle and crust, suggesting they were added after Earth was made. And new research suggests that Theia remnants may be even more prominent than scattered precious metals.

To understand how this could work, we need to go back to Wegener's continental plates and the realization that Earth's crust is a changeling.

As Earth 2.0 and the newly formed Moon began to take shape, in the centuries after the synestia, the two worlds followed the law of entropy and began cooling off. They remained entirely molten for a half billion years or so, which we know because of Neil Armstrong's floating anorthosite rocks. In Earth's early days, heavy elements like iron and nickel separated from lighter ones and sank through the molten mantle, forming Earth's dense core. Today, the mantle is hot and soft, but not quite molten. Earth's outer mantle behaves more like road tar or candle wax than a river of lava or a chunk of stone. But the candle-wax mantle is not uniform. Within it are two continent-sized blobs of rock that seem to be denser and chemically different from the rest of Earth's interior. These blobs, thousands of miles wide and as much as six hundred miles deep, stretch from the core-mantle boundary, bracketing Earth's core like a pair of earmuffs. One sits under Africa and one sits under the Pacific.

Geologists have debated the nature of these odd magma structures, which are known as large low-shear-velocity provinces, for years. Some people aren't convinced the blobs even exist. Evidence for these masses comes from seismic waves, of the kind that ripple through the planet during an earthquake. The earthquake waves slow down abruptly when they reach the blobs, which suggests they are made of different material than the rest of the mantle. Some geologists think these are slab graveyards—the remains of previous continental plates, subducted eons ago as Earth recycled itself. But other geologists have traced magma plumes from the blobs all the way up to the islands of Samoa and Iceland. Lava on those islands contains isotopes of elements that must have formed during the first hundred million years of Earth's history—potentially too early to be recycled pieces of tectonic plates.

Some people have wondered, mostly in hallways at scientific meetings, whether the blobs came from somewhere else entirely.

In 2021, a geodynamics graduate student named Qian Yuan suggested that they are from Theia.[11] According to this model, after the collision, Theia's core would have merged with Earth's—either inside the synestia or as the naked cores coalesced—and pieces of Theia's mantle, if it were denser than Earth's, would have remained distinct inside Earth's innards, like clumps in undermixed pancake batter.

To test this idea, scientists could go back to the isotopes and look for similarities between the Samoa and Iceland rocks as well as rocks from the Moon's own mantle. But none of the Apollo samples captured the Moon's interior. The astronauts dug up what they could, but they mostly sampled the rocks that had floated—the anorthosites—and rocks that had been liberated in near-side asteroid strikes, just as Wegener predicted.

To get to the Moon's deepest heart, you would need to sample the largest impact basin on its surface. And that is on the far side, near its south pole, where no human has walked yet. In 2019, a zippy Chinese rover called *Yutu-2*, "Jade Rabbit," landed on the eastern floor of a crater in the Moon's gigantic far-side scar, called the South Pole–Aitken Basin. It captured some of the first hard evidence of the far side's strange geology. *Yutu-2* deployed lunar-penetrating radar to peer inside the Moon, and found that the soil is thicker than scientists expected. The regolith goes about 130 feet deep before there's any sign of lava beds, or maria; scientists expected about one-fourth of that. The basalt surface has been buried by regolith, gardened through the eons by a constant aster-oid pummeling.

The rover's findings marked the first time humans have been able to study one of the Moon's most mysterious characteristics: the distinct difference between its near side and its far side.[12]

Until scientists saw the photos from early Soviet spacecraft and from Apollo 8, no one had any reason to think the hemispheres of the Moon would appear different from each other. But the surface of the Moon's far side is totally unlike the side we see from Earth.*

IN THE FIRST few years after Apollo, some people thought Earth acted as a shield, safeguarding the side of the Moon that faces us and leaving the far side exposed to whatever the solar system might throw at it. But most scientists now think the horrific circumstances of Earth and the Moon's mutual formation led to the discrepancy between the near side and far side. Once the Moon and Earth had cooled into separate bodies, perhaps when their synestia calmed down, the Moon quickly assumed a tidally locked position.

"Tidal locking" is a boring term for the graceful way the Moon main-tains the same face toward us. The Moon actually does rotate, but its spin is equal to the time it takes for the Moon to circle Earth. So it ap-pears to be locked in place.

Because the Moon's far side was more distant from Earth from the beginning, the theory goes, it cooled faster, forming a thicker crust. The boiling Earth, radiating warmth at more than 2,500 degrees Celsius, kept the Moon's near side warm. This produced a temperature gradient on the Moon, forming a crust atop a fluffy interior like a soufflé.

Aluminum and calcium would have condensed first on the cold side of the Moon. Thousands or even millions of years later, these elements eventually coalesced with silicates and oxygen in the Moon's mantle, forming anorthosite, those chalky white rocks that characterize the

---

* The far side is not its dark side, though this is so commonly misstated. As Kant said, the Moon always shows the same face to Earth because it is tidally locked to our world. But the far side is illuminated by the Sun just like the rest of the Moon. It's invisible to us, but it's certainly not night all the time.

Apollo haul. But the superheated near side formed a thinner crust. Over time, meteorite impacts would release vast flows of basalt from that thin crust, like driving a fork through a chocolate lava cake. The maria resulted from these flows.

Heat from the Moon's innards also played a role. The near side is full of KREEP rocks, an acronym for potassium (symbol K) enriched with rare earth elements, like erbium, europium, and other weird ones, and finally combined with phosphorus (symbol P). KREEP material has a lower melting point. It may have been the last stuff on the Moon to solidify, after the synestia turned to clumps, the Moon-wide ocean of lava froze over, the anorthosite rocks were locked into the white crust, and the Moon finally settled into the pearlescent orb it is today.

BUT NOBODY REALLY knows why the faces of the Moon are so distinct. No theory can explain it with the certainty of Wegener's craters or the giant impact.

Though Jack Schmitt tried valiantly to get NASA to land Apollo 17 on the Moon's far side, all six landings happened on its near side. Each landing site was a little different, but rocks from every single one included similar material. This is a troubling fact. The Apollo samples only reflect the locations where we touched down, not the whole Moon. They give us a history of a particular place, which, by extension, suggests that they can't offer us the history of the entire Moon or the entire Earth— neither version 1.0 nor 2.0. And that means the stories we have told based on those rocks are all incomplete.

The inconstant Moon, spectral symbol of impermanence, is still forcing us to reckon with our own lack of understanding. To unravel its entire history, we will need to go back to its surface and restore more pieces of it to the planet from which it came. To unravel Earth's history with the Moon we need to go back in time.

CHAPTER THREE

The Biographer of Earth

WHEN ANCIENT THINKERS considered the creation of Earth, they thought of it as the creation of everything there could possibly be. To them, there was no universe beyond the borders of the world. No one knew what the planets—the "five pacers" in ancient China,[1] "the wanderers" in ancient Greece—were like. No one had any inkling that these were entire worlds distinct from this one. With a few exceptions, before the age of telescopes most people could not see galaxies, which meant nobody knew what a galaxy was. Certainly no one knew about the supermassive black holes at the centers of the galaxies, nor the invisible cosmic glue known as dark matter that holds galaxy groups together. Earth plays the starring role in all the oldest creation tales. Everything else in the heavens took a supporting role.

Early stories cast the Moon and the stars as helpers on a greater quest. To Plato, the Moon's orbit was made for marking time as a way to comprehend eternity. "For this reason there came into being night and the day. . . ." Plato writes in *Timaeus;* "the month, complete when the Moon has been round her orbit and caught up with the Sun again; the year, complete when the Sun has been round his orbit."[2] The Sumerian Seven Tablets of Creation also describe the Moon as time's marker and the Sun's fellow; the Sumerian Moon God was the father of the Sun God. In the Book of Genesis, the Moon is created at the same time as the Sun as the "second light," the one that illuminates the night.

The scientific story of the creation of Earth is not altogether different, but the Moon plays a more vital role. Forged alongside Earth in the searing hot synestia, the Moon is more sibling than subordinate. Theia is really the mother of both Earth and the Moon. And our silvery sister is still a part of us, Earth's biographer, its first chronicler, and its most thorough accountant. There is no story about Earth that can exclude its influence, and there is no story about the Moon that does not tell us something about Earth.

Since the beginning of evolutionary time, the Moon has sculpted life on this planet. The Moon stabilizes Earth's tilt toward the Sun, making the Moon the captain of our seasons. The consistency of this tilt over millennia stabilizes our climate in turn. Life in all its endless forms, from corals to plants to humans, responds to the Moon's cues. Oxygen exhaled by these breathing organisms streams out of our planet's atmosphere, flies with the solar wind, and pools on the Moon as proof that we are here.

The Moon's unusually large size and its distance from Earth mean that Earth and the Moon are a system, working together. The Moon is not just a small satellite orbiting a larger world; in fact, the Moon doesn't technically rotate around Earth's center of gravity. Rather, both Earth and the Moon orbit around their shared barycenter. This is a bland term for a lovely concept: The barycenter represents the nexus of the pair's relationship. Earth and the Moon orbit their *shared* center of gravity. The barycenter is found not at Earth's core, but at an average of three thousand miles from its center. This nexus is why the tide bulges both on the side nearest to the Moon and on the side opposite it. Gravity pulls Earth and Moon together, but they are also pushed apart by the centrifugal force as they orbit their shared barycenter. The Moon's tug on Earth pulls water toward itself, and the centrifugal force also pulls water in the opposite direction, creating two bulges, and two high and low tides each day as Earth rotates.

All of these relationships are more than just a sign of kinship. The Moon may be necessary for all the things that make Earth unique, especially its defining characteristic: us.

○ ˙

EARTH SPINS ON its axis; tilts on its axis with respect to its orbital path; and points in a different direction over time. Every one of these traits is influenced by the Moon, and every one of them changes Earth and everything on it.

First there is night and day, which we have because Earth spins around as though it's mounted on an invisible spindle that runs from south to north. The equator sits perpendicular to that spindle, circling Earth's middle. Plato figured this out twenty-three centuries ago: "The Earth, our foster mother, winding as she does about the axis of the universe, [the creator] devised to be the guardian and maker of night and day, and first and oldest of the gods born within the heaven," he writes in *Timaeus*.[3]

Any classroom globe will show you the poles and the equator. If you don't have a globe around, you can make a model yourself.*

We all know the day lasts twenty-four hours—at least right now—because this is how long it takes for a given spot on Earth to rotate under the Sun. But Earth also experiences a lunar day, or the "tidal day," which is the time it takes a specific spot on Earth to complete one rotation under the Moon. A lunar day is fifty minutes longer than a solar day, because the Moon is revolving around Earth in the same direction that Earth is rotating. It takes a given location on Earth almost an hour a day to catch up to the Moon.

As Earth spins, some of its energy is transferred, via friction in the oceans, to the tidal bulges. And because it is spinning faster than the Moon orbits, Earth forces the high tide to occur ahead of where the Moon is located in the sky, instead of directly beneath it.

The bulges therefore sit slightly ahead of the Moon. And through complex interactions involving gravity and other forces, this sloshing water transfers energy to the Moon, pushing it into a higher orbit. The upshot is twofold: The Moon is moving away, and the rate of Earth's spin is slowing down, just as George Darwin figured out. This happens slowly, just 1.8 milliseconds every century, but over geologic time that adds up to a lot. The Moon has changed the length of our day by twelve hours since primitive life emerged on Earth. Fossil corals show that in the Silu-

---

* Go get a pencil, a piece of paper, and a pair of scissors. Draw a circle on the paper, cut it out as best you can, and find its center. Stick the pencil through the center. Now you have a disk of paper impaled with a pencil. The paper represents Earth's equator, and the pencil is the axis. It's more fun if you pretend the eraser is the North Pole. Twist the pencil between your fingers to spin between night and day, and remember that the Moon is responsible for the rate of its spin.

rian period about 430 million years ago, when the first bony fish evolved, Earth rotated on its axis every twenty-one hours, for a twenty-one-hour day. That means the planet spun 420 times for every revolution around the Sun. By the Devonian period between 419 and 359 million years ago—the time when fish developed lungs and learned to walk on land— the day was a couple hours longer. In another 200 million years, a day will last twenty-five hours, more like a day on Mars, and a solar year will last only 350 days.

THE MOON SHAPES the length of our days, and it plays a major role in what they feel like outside. Its gravity helps safeguard our tilted planet against climate chaos.

When the tumultuous early nebula organized into the Sun and planets, most objects settled along a flat plane with the Sun at the center. This invisible plane is called the ecliptic. The planets go around the Sun on the ecliptic plane as if they were tracing the grooves on a vinyl record. Sometimes you can see this phenomenon with the naked eye. At certain times of year, on a clear night, you can spot Venus, Mars, Jupiter, and Saturn arranged in a diagonal line. Occasionally, when you're lucky—or unlucky, depending on which astrology you believe, and whose history you are reading* —the planets even appear to huddle close.

---

* In May 1059 B.C.E., the southern sky held a supremely rare sight: Mercury, Venus, Mars, Jupiter, and Saturn, all grouped together within about seven degrees of sky. For comparison, your fist held out at arm's length covers about ten degrees of sky. It would have been a spectacular sight, viewed from anywhere on Earth. In ancient China, it would have been drenched with significance.

The five pacers, or wu bu (五步), were also known as the "Minister-Regulators" of the Supernal Lord, Shangdi. He was the supreme deity during the reign of the Shang, and he controlled the harvest, the weather, and victory in battle.

In May 1059 B.C.E., his pacers were flying in the beak of Vermilion Bird, the constellation that represents the southern sky. It is a fiery red bird, the Phoenix in Western lore, and one of the four major Chinese mythological symbols. The Vermilion Bird also represented the fortunes of the Zhou dynasty. The pacers in that constellation, coupled with a blood-red eclipsed Moon almost exactly seven years earlier, seemed to the vassal king Wen of Zhou like a clear mandate to seize power. He built a new calendar and inaugurated it in 1058 B.C.E.: the First Year of the Receipt of Heaven's Mandate.

The Mandate held that rebellion against an unjust ruler is not only a right, but consecrated by heaven itself. The Mandate guided Chinese politics for the next three millennia.

Earth is tilted with respect to this invisible plane, however, by about 23.4 degrees. As any elementary school student learns, this essential feature of our world gives us the seasons. Pick up a pencil and grasp it like you're about to write your name. The barrel of your pencil rests between your thumb and forefinger, and as the tip touches the surface, the North Pole is tilted, just like Earth.

In the Northern Hemisphere, winter arrives when the northern half of Earth is tilted away from the Sun, which appears weaker and lower in the sky. When it's winter in the North, the Southern Hemisphere is pointed closer to the Sun and it's summer down there. And the reverse is true: In northern spring, Earth's journey around the Sun tips the northern half closer to our star, and things in North America and Europe warm up.

Most of the planets are tilted on their axes, some by a tad and some by a lot. Venus is tilted by 3 degrees and doesn't really have seasons; it also spins backward. Uranus is knocked practically sideways, with a tilt of 97.7 degrees. Mars is tilted by about 25 degrees. Although Mars is a lot like Earth—similar axial tilt, a day (called a sol) that's just thirty-nine minutes longer than ours—it doesn't have much in the way of moons. Its twin satellites Phobos and Deimos may be captured asteroids, or may have been born of Mars the way the Moon was from Earth, but whatever their origin, they're dinky little spuds that play no major role in Mars's story. And the lack of a massive moon may be one reason Mars has such an intense climate.

In the past ten to twenty million years, with Jupiter and Saturn exerting their influence and no big moon to offer stability, the Martian axis has wobbled wildly between fourteen degrees and forty-eight degrees. When it's tilted more deeply, Mars's carbon dioxide polar ice caps point nearer to the Sun and can completely melt or sublime, turning directly from ice into vapor. This injects more carbon dioxide into the Martian atmosphere, causing extreme climate shifts.

Earth's axis wobbles, too, but not by very much. It remains remarkably stable over millions of years because the Moon safeguards it. During the past ten million years, Earth's axial tilt has shifted by only two degrees.

In 1993, Jacques Laskar and his colleagues at the Paris Observatory found that without the Moon, gravitational interference from Jupiter would push Earth around like a playground bully.[4] Earth's axis would tilt somewhere between zero degrees—ramrod straight—and a vertiginous eighty-five degrees. In such a scenario, during one millennium, Earth's poles would point almost directly at the Sun, and the equator would be frigid; then a couple million years later, everything would reverse. Imagine the snowy Antarctic at today's equator, and the tropics perpendicular to the Sun, frigid and dark. Such a wild wobble would make it hard for any life to survive for very long, especially large, land-dwelling, insatiable creatures like humans.

Astrobiologists have reexamined Laskar's claim in the years since, and some have argued that Earth would wobble between ten and fifty degrees, not quite as extreme a difference as Laskar predicted but one that's still horrifying to imagine. Past climates can provide a glimpse of what those changes would look like and how dramatically they could affect Earth. While Earth's tilt is not thought to be solely responsible for the rise and fall of ice ages, the changing angle and intensity of sunlight can trigger feedbacks that contribute to those global temperature shifts—and that's with just a tiny change in angle. One example is an interglacial cold period known as the Younger Dryas. When it came to a close about 11,500 years ago, global temperatures shot up—in Greenland, temperatures climbed by as much as eighteen degrees Fahrenheit in just a few decades, according to ice-core studies. Not long after the warm-up, humans living in the Fertile Crescent started to stay put rather than travel in hunting bands, and began to create civilization. What ended the Younger Dryas cold period is still disputed, but some paleoclimatologists think rapidly melting ice sheets shifted heat circulation in the oceans, especially the North Atlantic. The tilting of Earth's poles toward the Sun could have played a role in melting all that ice. What's clear is that the changes in Earth's climate would certainly have been more extreme if the Moon were not here to stabilize it. Today Earth is tilted at 23.4 degrees, but the Moon is making this angle shallower. Earth will reach its minimum tilt of 22.1 degrees about 9,800 years from now.

THOUGH THE MOON smooths out Earth's tilt over time, it does make our planet wobble on its axis, just a bit. This happens because of tidal forces caused by the Moon and the Sun.

The first person to figure this out was a Greek astronomer named Hipparchus of Nicaea. As a boy in what is now northwestern Turkey, he compiled records of local weather, and in adulthood he studied eclipses to measure the distance between Earth and the Moon. He came quite close to the real distance, a remarkable feat for someone living between 190 and 120 B.C.E.* Hipparchus is credited with inventing trigonometry, and his ideas were incorporated into the astrolabe, which was used around the world for more than seventeen centuries to calculate astronomical positions. Hipparchus's original works were lost to time; we only know of his work through later astronomers, most notably Ptolemy. But in 2022, scholars studying an ancient Greek palimpsest, a parchment that had been erased and reused, found a paragraph of his original star catalog covered up by a Christian codex.[5] It shows part of Hipparchus's attempt to map the entire night sky. He was the first to unify the vast astronomical heritage of the earliest Mesopotamian astronomers with later geometrical models, developed by his Greek forebears. Before Hipparchus, Western classical antiquity had two main types of skywatchers: the superstitious, Moon-tracking Babylonians and the cerebral, perfection-obsessed Greeks. The cosmos was like a pile of blank puzzle pieces with no images on them to help anyone figure out how they fit together.

Hipparchus took the puzzle pieces—the Greek idea of moving spheres and perfect realms—and painted them with the stars of Babylon. Scholars think he was the first to define stars' positions in the sky using two coordinates, similar to latitude and longitude, so they could be located apart from describing their position relative to constellations. His famous celestial catalog, completed around 129 B.C.E., used Greek knowledge of the motions of the stars, planets, Sun, and Moon to

---

* Hipparchus measured solar eclipses and Earth's shadow during a lunar eclipse, using trigonometry to calculate that the average distance between Earth and the Moon was sixty-three times the radius of Earth. It's actually sixty Earth radii, or roughly 238,000 miles. Not bad.

figure out where these objects were located and where they would be in the future. He then compared his own observations of their locations with the centuries-long records of Mesopotamian sky priests. And he realized that the stars as he observed them had shifted from earlier Babylonian measurements. The backdrop of stars was changing not because the stars themselves had moved over the ages, but because Earth was moving beneath them. This motion, now called precession, occurs because Earth wobbles on its rotation axis like a spinning top.*

To understand why this matters, we have to talk about the year.

Earth spins 365 times before it comes back to the same position relative to the Sun, which makes a tropical year. Westerners mark the beginning of this journey on January 1, but the tropical year technically stretches from equinox to equinox, the days twice a year when Earth's axis is tilted neither toward nor away from the Sun, resulting in day and night of nearly equal length. Astronomers—and everyone on Earth before the modern era—also measure the "sidereal" year, which marks the time it takes for certain stars to return to the exact same spot in the sky.

Because the Moon makes Earth's axis wobble, those certain stars actually do not return to the exact same spot over the centuries. Earth's axis over time points a slightly different direction in space. Eventually, even a trusty guide star like the North Star will seem to move out of place. Right now the North Star, a celestial point that appears to stay fixed throughout the year and thus helps with navigation, is the star we call Polaris. But during the Younger Dryas, humans in the Northern Hemisphere saw a different lodestar. And in another twelve thousand years or so, our North Star will be Vega instead.

Other stars seem to move even more rapidly. And that means the sky looks subtly different at the same time each year. Over centuries, the night sky can look quite different. This was a problem for ancient as-

---

* The pencil-and-paper trick can show you (roughly) what axial precession looks like. Grab the pencil and draw a circle, watching as the North Pole–eraser traces out the path. If you could trace your circle over the course of 26,000 years, this demonstration of Earth's axial change would be more accurate, but fortunately for both of us you don't have to do that to understand the point. Earth's axis completes one full rotation, one circle, every 25,772 years.

tronomers through the ages, who used the first visibility of stars, or the location of certain stars compared to the Sun and Moon, in order to mark events during the year. The first dawn sighting of the bright star Sirius, for instance, marked the start of the ancient Egyptian calendar year, because in 3000 B.C.E., Sirius rose at dawn in early July, when the Nile begins its annual flood. How does a civilization tell time over long periods when the timekeeping system itself is always changing?

WHAT'S MORE, the Sun's course across the sky is also out of sync with the Moon's. A lunar month represents the time it takes the Moon to return to the same alignment of Earth, Sun, and Moon, a lineup called syzygy. This lasts 29.53 days. The lunar month begins on the day of the new Moon, when it is invisible. The Romans called this day the "Calends," the word that gives us our modern "calendar." The eighth day falls at first quarter, a waxing half-Moon. The fifteenth day, which the Romans called the "Ides," is the day of the full Moon, and so on.

After twelve of these cycles, the Moon returns to the same location in the sky, hanging just about where it did on the first cycle. That adds up to twelve months in a lunar year, what's called a synodic year. But there are only 354 days in this year. If you are sitting on Earth, 365 days will go by before the Sun returns to the same place in the sky. This eleven-day gap poses a problem; after just three years, the lunar months are out of sync with the Sun by about a month.

Calendar makers figured out many creative ways to synchronize these two types of time.

The Maya civilization, who lived in what is now parts of Mexico and Central America, had one of the most complex solutions. They created three separate, interlocking calendars: Their Tzolk'in calendar runs 260 days and combines a cycle of thirteen numbers matched with 20 named days. The Haab' runs 365 days, counting eighteen months of 20 days apiece, and one 5-day intercalary month. Finally, their Long Count calendar keeps track of the years, including the time that has passed since the mythical starting date of the Maya creation, August 11, 3114 B.C.E. The Moon, captain of time on this planet, gave us the notion of calendars; it also means each of them is flawed, because its presence interferes so thoroughly with our planet.

THE WOBBLE OF Earth's axis interferes with more than our timekeeping systems. While the Moon stabilizes the axis to keep Earth's climate temperate over millennia, even a subtle wobble can cause dramatic climate changes on shorter timescales. First, it helps to understand something that bedeviled the marines at Tarawa: Neither Earth nor the Moon orbits in a circular path, but rather in an ellipse. They each have points where they are closest to and farthest from the bodies they orbit. The Moon was at apogee, its most distant point from Earth, on Tarawa's D-day. Every year, Earth also travels to its nearest and farthest spots around the Sun, called perihelion and aphelion. But Jupiter and Saturn stretch Earth's orbit so that it is sometimes almost circular and sometimes quite elliptical. Right now, Earth's orbit is near its most circular trajectory and due to become more oblong, on a cycle that spans about one hundred thousand years.

The Moon is not powerful enough to change the shape of Earth's orbit. But it does change the season when Earth experiences perihelion and aphelion. In the 2020s, Earth is at its closest point to the Sun in January every year, but ten thousand years ago, perihelion occurred in the Northern Hemisphere's summer. Rains deluged the Sahara during this time, known as the African Humid Period. The end of the rains shifted migration patterns in northern Africa, planting the first seeds of civilization.

THE EARTH-MOON SYSTEM'S orbital oddities—Earth's tilted axis, the precession of the equinoxes, and Earth's elliptical orbit—add up to a difference in the amount of sunlight that pours onto our planet through time. These changes are known as the Milankovitch cycles, for the Serbian scientist Milutin Milanković, who laid them all out a century ago. They contribute to cyclical periods of change in our planet's climate (and are a favorite contrivance of people who pretend humans are not causing global warming).

What matters most in these cycles is the Moon. It is so big relative to Earth, and it bears so much of the angular momentum of the Earth-Moon system, that it safeguards us from chaos. And it is just the right size to be of help and not harm. If the Moon were just 10 percent larger, the Moon itself would cause Earth's axis to become unstable.[6] If Theia and Earth 1.0 had been remade into a slightly bigger Moon and a slightly littler Earth, we probably wouldn't be here.

All this particularity has led more than a few scientists to speculate on the "anthropic principle," the notion that the universe is observable because it is organized in such a way that we arose and are still safely here to observe it. There are plenty of planets and plenty of moons out there, but only systems that are just right would give rise to creatures that can observe reality. Since we exist and are observers of the universe, then Earth must have those just-right properties, even if those properties are apparently rare. Only half of the Sun-type stars we've found so far could even host a planet of roughly Earth's mass within this "just right" so-called Goldilocks zone, where water can remain liquid, and so far, we have not found life on any of them.

THE MOON'S GRAVITATIONAL pull might explain another strange facet of our planet: our magnetic field.

As the Moon orbits in its elliptical path, its tug on Earth waxes and wanes with time. Just as the Moon controls the ocean tide, it pulls and stretches Earth's mantle, which encases the planet's liquid outer core like an egg white surrounding an iron yolk. Rocks feel solid, but they contain slight imperfections and cracks, which can lend them elastic tendencies. When a rock is under some sort of external stress, like heat, water, or a tidal pull, the rock can slightly deform. Changes in a rock's shape are captured in a measurement called strain, which can tell scientists about a rock's strength and stiffness. In a study in 2019, German researchers using a seismic monitoring station in Chile observed seismic waves, akin to earthquake waves, move through the rocks in time with the lunar tide's rhythm.[7] The half-day oscillations corresponded to 12.42 and 12.56 hours. The 12.42-hour period precisely matches the lunar tide, while the longer oscillation matches the Moon's elliptical orbit. The same phenomenon that stranded the marines at Tarawa is visible in Earth's rocks.

The Earth-Moon-Sun rotational tidal engine pumps more than 3,700 billion watts of power into Earth. Some of that is lost in the atmosphere, some dissipates in the deep ocean, and some dissipates in the tides, helping to fling away the Moon. But some of that energy is unaccounted for. Geophysicists have claimed the excess energy may be continuously injected into Earth's outer core, feeding the mysterious magnetic dynamo that creates Earth's magnetic force field. If you're lucky you can even see it, on clear nights at extreme latitudes. Earth's magnetic shield produces the spectral electric-green ribbons of the aurora borealis and aurora australis. Birds can see it, too. Any compass can feel it. The magnetic field surrounds the planet and flows around it in the solar wind, like a curtain flapping in a breeze. The field emanates from Earth's outer core, which swirls around like liquid, generating an electric current. Heat rising from hot radioactive rocks and the planet's own rotation both serve to warm and stir the molten iron. (The Moon controls that rotation, too, don't forget.) But these forces are still not enough to keep Earth's core piping hot.

Earth is constantly losing heat into space, as is the Moon. But the inner core is still a scorching 10,800 degrees Fahrenheit, as hot as the surface of the Sun. The core-mantle boundary is about 6,900 degrees Fahrenheit (4,100 degrees Kelvin). Scientists have long held it must be cooling very slowly from within, but in 2016, geophysicists argued that instead, Earth maintains the heat left over from the day Theia and Earth 1.0 coalesced into the Moon and Earth 2.0.[8] Today's warmth is a lingering gift of radioactive isotopes and the Moon. The Moon's tidal action stirs the core and mantle as well as the seas, and may provide the power required to keep Earth's interior sizzling.

The Moon may have shaped Earth's magnetic field from the moment of its creation. Even if the remains of Theia are not interred within Earth today, the doomed world would have reshuffled Earth 1.0's mantle, which might have altered its convection process, which drives Alfred Wegener's plate tectonics. Maybe, some geophysicists think, the Moon and its singularly violent creation gave us the uniquely active, churning, rocky, force-field-protected world we have today. Maybe Wegener's ideas about plate tectonics were Moon-related after all.

○˙

THE SWIFTLY ROTATING, tilted world we all inhabit has been remade countless times. Through the action of plate tectonics, Earth has erased prior iterations of itself. But we can look to the Moon to understand what early Earth, and the rest of the solar system, might have been like before these transformations. The Moon records an ancient battering that it must have shared with Earth, and in telling that story, it offers us a biography of the entire solar system. Understanding the nature of the early Moon has major ramifications for some of the biggest questions in science: What was early Earth like? What about the other rocky planets? How did life arise on Earth, and when?

The Apollo missions brought us a new understanding of not only the Moon and Earth's shared creation story but also what happened afterward. As scientists studied isotopes of Moon rocks, they noticed that the Moon's surface had heated up catastrophically, practically melting the entire thing, about 3.9 billion years ago. That's half a billion years after the Moon was made. Based largely on these rock signatures, scientists came to believe that 500 million years after the synestia formed Earth and the Moon, the Moon was blasted within an inch of its life by a hard rain of asteroids the size of cities. They would have liquefied the Moon—and, because it was right next door, Earth as well. In 1973, planetary scientist Fouad Tera and his colleagues termed this epochal bombardment a "terminal lunar cataclysm," which also became known as the Late Heavy Bombardment.

"It must in any event have been quite a show from the Earth," they wrote, "assuming you had a really good bunker to watch from."[9]

The story of the bombardment made a lot of sense. Thanks to Wegener, scientists knew the Moon was often pelted with asteroids. They also knew that some of the molecular signatures in the Apollo rocks can only be made in the most horrifying conditions imaginable, as if the Moon's entire mantle had been liquefied and rolled inside out. The Moon had, indisputably, suffered some kind of calamity.

Earth, being larger than the Moon, would have borne the brunt of this beating. For a half century, scientists believed this late bombardment kept our planet molten and sterile for its first billion years, give or take a few hundred million. Not until 3.8 billion years ago, when the assault ceased, could life have taken hold. We can't see the scars of this drubbing because of plate tectonics, but that doesn't mean it didn't happen. The question since the Apollo missions has been when it happened, and for how long.

But then, in 2009, things changed. A new satellite called the Lunar Reconnaissance Orbiter (LRO) arrived at the Moon and started taking detailed photos of the Moon's mottled face. On March 17, 2013, NASA scientists who study the satellite's images saw a bright flash in Mare Imbrium, the "Sea of Rains," one of the Moon's largest basins. After the flash, LRO spotted a brand-new, sixty-foot-wide impact crater. Emerson Speyerer and his colleagues at Arizona State University teamed up to start looking for more of these new impact basins, and compared fifteen thousand images from the LRO at different times, but with the same lunar areas in the frame.[10] The before-and-after imagery showed 222 new craters and more than 47,000 new splotches, which occur when a meteorite impact forms a crater and splats Moon dust around, just like Wegener's cement powder. Not only did the NASA probe find that the Moon's craters are newer than anyone thought, but also that Mare Imbrium, one of the most iconic of these, might be much bigger than anyone thought.

The LRO camera teams found rays of debris extending out from Mare Imbrium and scattered all over nearby basins. This was a troubling sign. It suggests that whatever whacked into the Moon and formed Imbrium probably scattered Moon bits throughout nearby impact areas. It means we can't say for sure where any lunar rock originates, and in what cataclysm it was brought to the Moon's surface.

The implications are huge. The Apollo samples, priceless Moon bits locked in NASA's lunar sample clean room, might all be biased toward one singularly horrific impact. Mare Imbrium might be fooling us. One of the trickiest samples is troctolite 76535, the oldest and most interesting piece of the Moon brought home to Earth.

During Apollo 17, Jack Schmitt picked up troctolite 76535 in a region called the Sculptured Hills, which geologists thought would contain rocks from the Sea of Serenity. But there's a chance it contains rocks from the Mare Imbrium impact instead, or maybe even the impact that excavated the solar system's biggest crater, the Moon's South Pole–Aitken Basin. Troctolite 76535 is now one of the most controversial rocks to come home from the Moon. Its chemical makeup shows that it must have been formed at great depth, when the Moon was warm. It must also be about 4.25 billion years old, which means it is a chunk of the primordial Moon. And the newest studies of this rock show that only a truly terrific collision could have mined it from the Moon's belly. Whatever troctolite 76535's provenance, there's a distinct possibility that many of the Apollo samples were liberated from the Moon's innards after one or two huge, singularly destructive impacts. The Moon may not have suffered an unusually late asteroid blitz at all. And that would mean Earth didn't, either.

Today, the story of the Late Heavy Bombardment is in question, with many planetary scientists wondering whether it ever happened. If it didn't, if early Earth was warm and calm and Edenic, then there's no reason for life to have waited a billion years to take hold. Maybe life showed up right away, and simply stayed boring for longer than we thought before evolving into ferns and dinosaurs and us. Or perhaps the young Sun played some role in life's earliest days. The Moon might be able to answer that question, too.

The Sun's rate of rotation during that first boring billion years of Earth's existence would have affected the timing and frequency of solar outbursts, like flares and coronal mass ejections, clouds of radiation the Sun flings toward the planets at random. These emanations could have influenced the evolution of life. Perhaps they introduced a spark of radiation into Darwin's warm little pools, or ripped and knitted together the proteins that eventually built DNA, or sterilized any early attempt at cell building or replication. Understanding the Sun's spin history is hugely important for understanding the origins of life on Earth. Lunar dust contains a record of that history.

If the baby Sun rotated slower than comparable baby stars, its leisurely pace and placid energy could have helped early life limp along. But if the baby Sun was a super-spinner like most of its brethren, it would have erupted in solar flares ten times a day, scorching Earth and even whisking away its atmosphere. Moon dust contains evidence for the former theory. Elements like sodium and potassium are less abundant on the Moon than on Earth, and it turns out that a slow-spinning baby Sun would have rotated at just the right speed to shear away the missing amount of the Moon's material, according to research from 2019.[11] The young Sun was a slow spinner, Earth was big enough for its gravity to hold on to its material, and now here we are.

One controversial Apollo 14 sample, a rock nicknamed "Big Bertha," contains chemical signatures that look more Earthly than lunar. Does the Moon harbor bits of Earth, blasted from our planet—by some titanic impact, such as the one that killed off most dinosaurs, or some other collision that has been lost to tectonics and time? Scientists have found dozens of Moon rocks that returned home to Earth as meteorites, which landed in Antarctica, in the Sahara, and in other austere locations. It would not be terribly surprising to find similar Earth meteorites scattered on the Moon. Even if Big Bertha is not from Earth and is instead a chunk of the Moon, it still tells us something interesting, much like the now-controversial Late Heavy Bombardment. The Moon rocks are not monolithic. They have plenty of new secrets to tell.

$$\mathrm{O}^{\textbf{·}}$$

BIG BERTHA MAY or may not be from our planet, but scientists agree that the Moon is host to another Earthly souvenir: oxygen. This element of breath pools on the Moon for the same reason that we have full Moons, new Moons, and eclipses: because Earth, the Moon, and the Sun occasionally line up.

Arguably the clearest proof of life, oxygen is a volatile element and the most important component of the air we breathe. It is heavy, as gases go, and it can stay in the atmosphere for a while rather than escaping to space the way hydrogen does. Oxygen is highly reactive, which you know if you've ever lit a campfire. It reacts with methane, other atmospheric gases, and minerals inside Earth's rocks. Left to its own devices, Earth would suck oxygen out of the atmosphere and bury it in the planet's crust and eventually its mantle, where it would remain, forever. The fact that it does not do this, and that Earth's air is made of so much oxygen, is an everlasting gift of plant life. Plants respire, consuming carbon dioxide and pumping oxygen into our atmosphere, which is about one-fifth oxygen in total. The solar wind—that quirk of our modern rotating star—strips some of it away. When the Moon and Earth are lined up, which happens about five days per month, the Sun blows atoms out of our atmosphere and carries them all the way to the Moon, where most of the oxygen is locked in the regolith and some remains in the Moon's aura, its tenuous exosphere.

Oxygen holds an important lesson for us as our search for signs of life on other planets becomes increasingly sophisticated. Powerful telescopes may be able to spot it in the aura of distant exomoons, hinting at life on other planets around other stars.

Astronomers are still not certain they've really seen an exomoon, which is just to say a moon that orbits a planet that is not in this solar system. Extrasolar planets are abundant, which is one of those bits of information, now common knowledge, that was unthinkable not so long ago. The Kepler Space Telescope, named for Johannes Kepler, found worlds aplenty when it stared at a small patch of sky between 2009 and 2018. The telescope's namesake could never have imagined it, but astronomers now think practically every star in the sky harbors planets.

If exoplanets are plentiful, and moons in this solar system are plentiful, then it's not a giant leap to imagine that exomoons are plentiful, too. In 2018, astronomers announced the first possible evidence of such a moon, orbiting around an exoplanet called Kepler 1625-b. David Kipping and Alex Teachey of Columbia University in New York used the Hubble Space Telescope to scrutinize the star Kepler 1625; the planet Kepler 1625-b; and the putative moon, whose existence wasn't certain enough to earn an Earth-given name. Orbital dynamics suggest that the planet is gigantic, several times the mass of Jupiter, and its moon is somewhere around the size of Neptune.[12] The exosystem is in one sense a supersized version of the Earth-Moon system. Since Kipping and Teachey's announcement—which they stopped short of calling a discovery—many astronomers have turned their own attention toward other exoplanets. As of this writing, no one has definitively confirmed an exomoon. But eventually, exomoons will likely be as commonly known and as unremarkable as the panoply of planets, scattered like dust motes throughout the cold and empty cosmos. Their presence will be mind-expanding, in the way that the moons of Jupiter were four hundred years ago, in the way the Apollo missions were in the 1960s, and in the way exoplanets were in the early 2010s. If there are any other worlds like this one, with continent-spanning gardens and life-filled oceans and thinking creatures that can reflect on it all, we may find that those worlds have their own biographers traveling alongside them, influencing them from their beginnings.

WHILE THE MOST notable relics of Apollo—the Moon rocks—are now back down on Earth, the astronauts left a few important things behind. Along with an American flag, a plaque, their footprints, and some trash, Neil Armstrong and Buzz Aldrin left behind a couple of science experiments. One of these was a two-foot-wide panel bedecked with one hundred mirrors, designed to return any light in exactly the direction from which it came.

Every other Apollo experiment eventually faded into history, from the original rock samples to the seismometers that measure moonquakes and other geologic activity. But fifty years after the mirrors were delivered, the retroreflector experiment is still going strong. Telescopes in Texas and France continue to use the retroreflectors every day. A telescope on the Calern Plateau in the south of France has been staring at them for a half century. The telescope's 1.5-meter mirror also contains a laser, which bombards the Moon with ten pulses of photons per second. Only a few will make the trip, and even fewer of those will come all the way home. The round-trip journey takes about 2.4 seconds, and in the echo of light that returns, astronomers can discern the distance between the Moon and Earth to within a few millimeters. Though this seems incredibly precise, it's somewhat less so than scientists would like, total accuracy being limited by Earth's atmospheric interference.

This is how astronomers learned that the Moon is spiraling away from Earth at a rate of about 3.8 centimeters, or 1.5 inches, per year. The change in the rate of this separation is roughly equivalent to the rate at which your fingernails grow. About six hundred million years from now, the Moon will be so distant that it will no longer eclipse the Sun.

About two billion years from now, when humans will likely no longer exist, the Moon will be too far away to stabilize Earth's tilt. Earth's axis will tip toward the Sun, and the unstable hellworld Laskar predicted will come to pass. Earth's climate will experience regular, possibly violent, shifts. Its tides will falter, and so will its rock tides—the Moon-related stress and strain in the planet's very innards. If life in any form is still here, the slow retreat of the Moon will very likely pose an existential threat.

BUT FOR NOW, at least, the Moon will continue to guide our lives. Theia made the Moon and its companion, and may have left parts of itself buried within our world. The Moon's motherworld may also have donated to us its nitrogen and more crucially its carbon, the element that enables our existence. After the Moon and Earth coalesced, lunar tides had a profound effect on our planet's geologic and evolutionary history. Extreme Moon-driven tides mixed the primitive oceans like a ladle stirring a pot full of soup, dredging nutrients from the bottom to support the food chain on which our primeval ancestors depended. Without the tides and their effect on ocean currents, nutrients might have remained on the seafloor, never to be used by the vast chain of marine life. Half of the energy required for ocean mixing is provided by the dissipation of tides on the ocean floor.[13] After the first organisms arose, the Moon probably kept time for the symphony of life, and it still may play a role in our own fertility, physiology, and behavior, in ways that surprise many modern scientists and in ways they do not fully understand.

Large-scale phenomena like the tides are obvious, especially to humans who think in pictures and stories on a scale familiar to us—changing seashores, a stranded invading military force. But the barely discernible phenomena at play between Earth and the Moon may be just as vital to life's rhythms, and may have been so from the very beginning. We would not be here without the Moon.

PART II

HOW THE MOON
MADE US

CHAPTER FOUR

The Moon and the Origin of Species

*Nature proceeds little by little from things lifeless to animal life in such
a way that it is impossible to determine the exact line of demarcation.*

—Aristotle, *History of Animals,* book 8, part 1

A FEW HUNDRED million years after the synestia, after Earth and the
Moon coalesced in their natal cloud, after they became spherical and
mostly solid worlds, they started to become fully themselves. The process of becoming a planet turns out to be surprisingly quick, especially
on geologic timescales. Even during Earth's Hadean eon, four billion
years ago, the planet looked a lot like home. Earth had oceans and landmasses, maybe even continents. Earth had clouds and rain made of
water, some probably delivered by comets and some left over from the
planet's birth inside Immanuel Kant's spinning protoplanetary disk.

The opaline Moon, just eighty-three thousand miles away, appeared
nearly three times larger in the sky than it does now. Earth's rotation
took only twelve hours, meaning a six-hour night and day and tides that
rolled in at a frenetic pace. But otherwise, Earth then and Earth now
looked much the same. Imagine: You walk toward the sea, on a shoreline
made of freshly baked rock—there is no sand yet, because there is no
life, much less life as complex as the seashells that form most white
sand. It is hot. Really hot. No ozone protects you from the Sun's radiation; instead it pours through the sulfurous atmosphere, which stinks of
rotten eggs. Foamy waves roll in and you look down, noticing a gap in

the rocks where a tide pool has formed. It's full of turbid brownish fluid. You dip in a finger and taste it. It tastes like soy sauce—that rounded, bitter, earthy, sour umami tang. This is the flavor of amino acids, like lysine, glycine, and glutamate. We now know that these compounds, the so-called building blocks of life, are present throughout the solar system, even permeating the tails of icy comets. But as far as we can tell, only once have they been pressed into service as life's raw materials.

The full story of life's origin is as murky as the brackish water in which it probably occurred, but what is increasingly clear is the Moon's vital role.

IN THE VAST, roiling oceans of primitive Earth, delicate chains of amino acids would be unlikely to connect in order to form life or to survive very long once they did—either in the stygian depths or on beaches. Tide pools gave them safe haven, however. As the Moon circled overhead, our soy-sauce pool filled up with warm, salty water, then dried out as the tide ebbed. It filled again as more water rolled in, and the cycle repeated. A cycle of hydration and dehydration produced larger and more complex molecules in a process called polymerization. Factories use the same basic phenomenon to make plastic.

Life may have begun in the deep, along a gash in Earth's crust where chemicals bubbled up into the superheated water, only to be sloshed to the surface by the Moon's pull. Or life may have begun in the shallows, eons before our backboned relatives learned to walk there. Since at least the days of Charles Darwin, many have reckoned that the first living things sparked into being in the liminal world of tidal pools.* After the first amino acid chains folded themselves into complex proteins (give or

---

* In a letter to a friend in 1871, Darwin wrote that life may have originated in "some warm little pond with all sorts of ammonia and phosphoric salts." By 1953, the American chemists Stanley Miller and Harold Urey showed this idea had merit. The duo threw sparks on a mixture of ammonia and methane and concocted a brown broth full of amino acids. These reactive little molecules contain carbon, hydrogen, nitrogen, and oxygen. When they link up in long chains, they form proteins, which provide the chassis for the machinery of life. Chemists repeated the famous Miller-Urey experiments in 2007, adding some other chemicals known to exist on early Earth, and replicated the production of amino acids. Though the recipe for life is still unknown to us, the basic ingredients could have been cooked up from scratch, using the atmosphere—and, crucially, the Moon's tides.

take a couple hundred million years), life oozed out of the muck. The first cells learned to make copies of themselves, and their copies made more copies. Just as those primitive forms began to replicate, the Moon yanked them out to sea and swirled them around in a cocktail of silt and nutrients. After life diversified into endless forms of bacteria, plants, and animals, the tides ensured that nothing remained at rest, constantly flowing, sinking, rising, mixing, evolving.

○ •

OURS IS A singular history. As far as we know, no other planet has ever given rise to the circus of eating, sleeping, breathing, and procreating life-forms that populate this one and endow it with air and oil, not to mention a global society that uses those things as an engine.

Life may have arisen early in the planet's history, soon after the simmering Earth and its incandescent, molten Moon cooled off. But it was agonizingly boring for the vast majority of the existence of the Earth-Moon system. The microbes that burst forth in Earth's oceans stayed there for an eternity, making copies of themselves but otherwise doing mostly nothing. Land-based life evolved in just the most recent 10 percent of Earth's existence.

Only once in the long, violent, phantasmagorical history of life did vertebrate animals leave the water to live on land. They did go back— our backboned relatives moved back to the water to evolve more complex forms,[1] and much later, mammals went back, too, transforming from diminutive creatures that look like dogs but are more closely related to hippos into charismatic dolphins and gargantuan whales. But as far as we know, fish only left the water once, to evolve into every vertebrate that has ever walked on Earth, from dinosaurs to elephants to us. Those fish probably came in with the tide, meaning that it was the Moon that dragged our piscine progenitors ashore.

COMPLEX LIFE WAS still fairly new when this happened back in the Devonian period, which stretched between 419 and 359 million years ago. This sounds like a distant epoch, but in the timescales of Earth and the Moon, it is a moment ago. More than 4 billion years had elapsed since the synestia and the Hadean eon.

Back then, when the day lasted only twenty-one hours,[2] the Moon loomed 5 percent larger in the sky, and its nearness produced sizable tides, much more extreme than those that wash across the world now.

Plants, maybe beached by tidewaters, finally began to colonize the land. The first trees and forests sprouted during the Devonian, also called the Age of Fishes. Although simple animals like comb jellies had been around for about 140 million years, new forms of fish were just beginning to emerge as the oceans churned out a kaleidoscopic array of new life. Some of these forms even began breathing the air.

Today's fish breathe by inhaling water and exchanging oxygen and carbon dioxide through filaments in their gills. You and I inhale air and exchange oxygen and carbon dioxide using alveoli in our lungs. But most modern fish have a version of lungs as well as gills. The lung-like organ is called a swim bladder, which acts like oxygen-filled ballast to keep the animal from sinking or floating up too far. Some fish, like the aptly named lungfish, can also use these organs to breathe air.

That these air-filled organs exist in all kinds of fish is evidence that lungs evolved long ago. They must have evolved before our early ancestors cleaved into two distinct groups of fishes, which happened some 443 to 420 million years ago. Lungs are one of the oldest tools in evolution's kit. But why would they have evolved at all, when marine animals can get the oxygen they need from the water?

During low tide, shallow pools at the tidal zone would get warmer and lose oxygen. A fish that was stranded by the tide's ebb would die within a few minutes if it couldn't make it back to the water. "In such an environment it might make sense for the fishes to evolve air-breathing organs," Per Ahlberg, a paleontologist at Uppsala University in Sweden and a leading figure in tetrapod evolution research, told me. If the ancestors of our inner fish could not yet walk, maybe they could at least breathe a little air and live to pass on their genes.

Sometime in the Devonian, these primitive lunged fish went two separate ways. These two main groups define the beginnings of Earth's animal kingdom. One group became the ray-finned fish, the ancestors of most familiar fish species, from goldfish to seahorses to salmon. The other group were the lobe-finned fish, which became the ancestors of every other backboned animal on this planet.

A few million years after lungs evolved, some lucky lobe-finned fish—fat, torpedo-shaped, probably greenish gray to better hide in the kelpy shallows—began to evolve another new trick. When the tide went out, the water level changed by as much as fifty feet. These fish would get stuck in shallow tide pools, but they found a way to escape.

As the water receded under the Moon's pull, ray-finned fish would flop on the drying sand, their gills flaring hopelessly. The fatter, larger, lobe-finned fish that dined on them would be unrecognizable in your family tree, but they are a key part of your lineage. Their fleshy front fins were sturdy enough to bear their weight. Inside them were bones resembling a humerus, an ulna, and a radius: the bones that make up your arms. And these armlike bones were attached to the body through a single joint on each side, the way your shoulders connect your humerus to your torso. With these advanced fins, our corpulent ancestors could move with purpose even without water to swim in. They could heave themselves across the sand, shimmy toward another shallow, silty refuge, and head back to the water. By breathing the air and moving their limbs, they would not be left to suffocate or starve until the Moon dragged the water back across the shore to set them free. Evolution favored the forms that could cover more ground: Flatter, stubbier, more evenly shaped fins could help a fish propel itself along the bottom of a shallow shoreline. Imagine walking in sand with shoes on versus walking barefoot—it's all in the traction. Over eons, the fins morphed into fleshy lobes that helped our distant relatives lift their heads and crawl. The Moon guided the strongest creatures to the shoreline, where they transmogrified into a dazzling array of shapes and sizes. The cradle of vertebrate diversity was the tidal zone.[3] And the strongest stayed in that area, too. The smallest, most delicate among the ray-finned fish moved away from land and toward the safety of the deep sea, where they were out of reach of their lobe-finned predators, and where the Moon had a harder time controlling their fates.

After many generations the fleshy lobe-finned fish developed random changes that helped them walk faster, and the cycle repeated: The Moon pulled the tide away; the fish, out of water, walked.

After a few hundred million more spins around the Sun, bipedal mammals descended from these fish chipped some of their bones out of rock exposed on Canada's frigid Ellesmere Island, near the Arctic Circle. The humans named one of these fish *Tiktaalik* and saw, in its fossilized remains, echoes of themselves and a link to our aquatic history. *Tiktaalik*, the first fish with a neck and a primitive wrist, likely had both lungs and gills. Fish like *Tiktaalik* and their relatives ushered in the tetrapods, the first four-legged beasts to walk on land. The descendants of the first fortunate walking fish, stranded by the tides, became the common ancestors of the dinosaurs, and the birds, and the reptiles, and the mammals, and me, and you.

THOUGH THE ROLE of the Moon in sculpting evolution makes intuitive sense, paleontologists didn't have good evidence for this relationship until very recently. In 2014, an Oxford astrophysicist named Steven Balbus analyzed the influence of the tides on the lives of Devonian fishes. Using lunar recession rates, he calculated that four hundred million years ago—just as tetrapods were emerging from the seas to walk on land—the Moon was about 5 percent closer to Earth.[4] Spring tides were stronger thanks to the closer Moon's greater pull, and would have repeatedly beached fish swimming in the intertidal zone. Such stranding events would have been even more likely in places where Earth's wrinkled surface features, like the shape of the coastline or the depth of the water, conspired to cause more extreme tides. The same coastal particularities shaped the dodging tide at Tarawa.

Hannah Byrne, an Irish oceanography graduate student, was captivated by the idea that the tide could have served as the tipping point that pushed our fish relations to walk. Byrne patiently walked me through her supercomputer simulations of tidal physics, which turns out to be much more complicated than water advancing and retreating. "It's all lies, what you have been told," she said, only half joking.

Water is the most powerfully erosive force on this planet, and, over time, tides weather continents. They sculpt continental shelves and build beaches. They change the shape of ocean basins, which can produce an effect known as resonance. Let's revisit the mechanism of the tide. As Earth spins, the Moon's gravity tugs on it, pulling the waters toward the Moon (and, because of centrifugal forces, also raising the waters on the side opposite the Moon). Earth's oceans experience two bulges daily. Most shores experience two high and two low tides a day, about 12.4 hours apart.* As all this water is sloshing around, waves are also moving, pushed by wind and the tide itself. A body of water can become tidally resonant, or synced up to the tide, when its dimensions match either one-fourth or one-half the length of the tidal wave. Put another way, this happens when the amount of time it takes a tidal bulge to travel into the basin is the same as the time between those high and low tides. When this takes place, the energy from the Moon's gravity matches the geography beneath the water. This alignment, like the Sun-Moon-Earth lineup of a full Moon, amplifies tidal energy even more and makes for extreme tidal shifts. Think of a person on a swing, rising and falling to the same height with every sweep of their legs. Now imagine someone else providing a push on their upswing. The swinger will go higher and faster next time. In a tidally resonant basin, Earth's landscape gives the ocean tide an extra push, causing a dramatic increase at high tide and a dramatic decrease at low tide. This is why northern Atlantic latitudes experience such dramatic tides, like in the Bay of Fundy, which separates Canada's New Brunswick and Nova Scotia provinces and touches the northeastern tip of Maine. The water there falls forty-three feet from high to low tide. As we've seen, a similar high-tide amplifier happens on the Normandy coast of France, and proved pivotal during World War II.

THE UPSHOT OF all this physics is that vast amounts of water are moving around on the surface of Earth all the time, in a regular cadence. A constantly changing water level means a constantly changing water temperature, salinity, and chemical mix, not to mention a changing shoreline as tidal flats are continually exposed and flooded. This flux has tremendous power over life. Byrne wanted to know whether the tide's power could push and pull life to shore.

She first met Balbus at a talk in 2014 while she was studying the tide's effect on nutrients and temperature on Pacific reefs. After hearing his

---

* This would be easier to understand if you could just cut the day in half, but remember the tide advances by about fifty minutes a day. The Moon moves quickly along its orbit—which is why it looks different every night. It therefore takes about fifty minutes each day for any given spot on Earth to catch up to the place it was the day before, directly opposite the Moon.

presentation, she switched her research to tide models, aiming to develop a quantifiable way to study his theory that the Moon influenced tetrapod evolution.

Working with Balbus and others, Byrne found that 425 million years ago, when air-breathing lungs began to evolve, the landmass that is now found in South China had the highest tides in the world. Later, in the Devonian period, this is also where fish learned to walk on land. These tremendous tidal changes provided an ideal environment for life in the water to move onto land.[5]

According to Byrne's research, these tides were so extreme because Pangea had not yet formed. Pangea was one of Earth's ancient supercontinents, a compendium of all of Earth's landmass smooshed together in one single structure, a now-accepted geologic truth that was first proposed by Alfred Wegener. Earth still had multiple ocean basins, increasing the number of spots where extreme tides could take place. At narrow points, tides on the proto-Pangea's shores would have been colossal, Byrne said. Byrne studied another major landmass where tetrapods are thought to have originated, part of an ancient supercontinent called Laurussia. This landmass is now found across central Europe, from the mountains of western Spain to the rocky outcrops of England's Northumberland. In the South China region and in the Laurussian, the shapes of ancestral ocean inlets, the tides from the Moon, and the geography formed the perfect conditions for beached fish to begin walking.

As these unfathomably ancient oceans began to close up, the supercontinent Pangea began coalescing. The crusts of these oceans have mostly been destroyed, but the aftermath of their demise is still visible. The closure of an Atlantic ancestor called the Rheic Ocean created the largest mountain-forming event in the Paleozoic. We call the result the Appalachian Mountains.

BYRNE USED GEOLOGIC evidence to develop ancient bathymetry, reconstructing the depth of the proto-Pangeaic continental shelves to find out how high the water would rise. In some areas, the tidal zone would only have been inundated in superhigh high tides, opposite the effect at Tarawa. The landmasses that hosted these ultra-high-tide zones are found in what is now China, northeastern Canada, the United States, Lithuania, Latvia, Estonia, and Iran.

Byrne's paleotidal data matches the fossil record remarkably well. The fishy forms emerging from their eons of stony sleep in these extreme tidal zones all reflect a transitional period between lobe-finned fish and true land walkers. "Something special was happening here," Byrne told me. The creatures that began dominating Earth late in the Age of Fishes were becoming land walkers and air breathers, just as we are.

○˙

ALL THOSE LIFE-FORMS, and all life everywhere on Earth, evolved according to a daily cycle between day and night. The Moon's inconstant light plays a key role in that cycle, and scientists have only recently begun to understand how.

Most life is diurnal, meaning it lives out its days and nights according to the spin of Earth toward and away from the Sun. The scientists who study biological rhythms and how they synchronize with this day-night cycle are called chronobiologists. Their primary focus is the circadian rhythm—from the Latin for "around the day." Every cell in your body has a molecular clock that sets the pace of your existence, from your metabolism to your breath to when you feel tired. A variation of this clock exists in every cell in every kingdom of life. For most creatures—fish, trees, frogs, you, me—the light of the Sun sets this internal clock. And our bodies don't just respond to the dawn or the dusk; the internal clock lets life *anticipate* these changes and adjust accordingly.

Our clocks can be set to a different phase, but it takes some time, and it generally requires a lot of sunlight in the morning and a lot of darkness at night. Anyone who's ever had jet lag has experienced the circadian clock going out of phase. Chronobiologists study how cells tell time, how organisms learn to live in time, and how our modern life—from long-distance travel to artificial night lighting to poor diet choices—can interfere with all these signals and disturb the clock.

But chronobiologists have recently begun to examine a different kind of rhythm. There is more than one light in the sky, and more than one way to set our internal clock. Life has a circa*lunar* clock as well. And scientists think it might be very common in the animal kingdom, because it was first found in marine animals, the ancestors of us all.

TELLING TIME BY sunlight and moonlight is rather like telling time both by the watch and by the calendar. The two are useful in different ways, as prehistoric humans would eventually figure out. The Sun provides a daily and a seasonal cue, and the Moon provides a monthly cue, both from its light and from its interaction with Earth's magnetic field and gravitation. When an organism uses both, it can sync up its rhythms, like its reproductive cycle, to the season, day, and even hour. This is true not just for animals but also for the leaf movements of plants. On Earth, on the International Space Station, and in constant darkness, plant leaf movements change in keeping with the rhythm of the Sun and Moon.[6]

It didn't take long for humans in antiquity to make this connection between life and the cycles of the Moon. Aristotle references the edible eggs of sea urchins in a text called *On the Parts of Animals,* written in 350 B.C.E. He recognized that the size of sea urchin eggs varied over the lunar month. "Though the ova are to be found in these animals even directly they are born, yet they acquire a greater size than usual at the time of full moon; not, as some think, because [sea-urchins] eat more at that season, but because the nights are then warmer, owing to the moonlight," he wrote.[7]

He was wrong about the temperatures, but he was right about the urchins responding to the moonlight. And sea urchins are far from alone. In the Arctic winter, when the Sun does not rise above the horizon and yet the full Moon is visible for four or more days, zooplankton sink and rise to the ocean surface en masse according to its light. Countless other marine species, from fish to corals, sync their mating, feeding, and spawning schedules with the phases of the Moon.

ON REEFS AROUND the world, from the Great Barrier Reef to the middle of the Red Sea, corals time their mating dance according to the full Moon's appearance.[8] Only after the full Moon has shone upon them will they release their pearlescent sperm and eggs, in a midnight phantasmagoria that biologist Oren Levy describes as "the greatest orgy on Earth."

Levy is an Israeli coral researcher who grows corals in tanks in his lab at Bar-Ilan University to study their spawning behavior and how it changes in response to light pollution, which interferes with the Moon's beacon. When he is not raising corals by hand, Levy snorkels to a reef in the Red Sea, near the Israeli resort town of Eilat on the Israel-Jordan border. It is the world's northernmost tropical reef, and corals there have been exposed to development, pollution, and artificial light for millennia. And yet they still use the Moon as their guide.

"We are talking about an organism that doesn't have any eyes. And it can still synchronize this behavior to the Moon's cycles," he told me.

The corals, which are tiny animals, produce parcels that wait like deliveries on a doorstep, near the threshold where their tiny bodies meet the sea. Then in an instant, in one of the most stunningly synchronous events on this planet, every coral releases its sperm and eggs. All at once, a pink blizzard of uncountable seeds floats up toward the light of the Moon. Many seeds will end their journeys as food for fish and other larger animals. But some coral sperm and eggs will combine, producing new coral larvae, which will bob with the tide until they can find a hard surface to anchor on and build a new city.

The seas' temperature, wind, and sunlight intensity set the month of spawning. But the Moon and its light set the day and the hour. Corals must release their packets at the same time to have any chance of forming new corals. This Moon-mediated mating dance may be more important than ever as corals worldwide succumb to mass bleaching events and other ravages of a changing climate. New generations of corals will need the Moon to colonize the reefs built by their ancestors. The next time you walk outside under a full Moon, witnessing the milky glow it casts over the trees and the grass and the buildings, think about what is happening, that very night, within this planet's oceans. How many organisms are being born, guided by the light of the silver pendulum in the sky.

For many organisms the Moon is a vital "zeitgeber" (a word borrowed from German because there was no English equivalent that means something like "time giver") just as much as the Sun is. But only very recently have chronobiologists started to unravel how this works. The ability to tell time by the Moon has genetic underpinnings, which probably date to the origin of genes, which is to say the origin of life.

In 2013, chronobiologists found in a marine worm and a sea louse the first evidence for a genetic "Moon clock" as distinct from a circadian clock.[9] The speckled sea louse, *Eurydice pulchra,* is a tiny crustacean one-third of an inch across and in the same taxon as crabs and lobsters. It lives in the intertidal zone and burrows into the sand when the tide goes out, turning itself black for protection against the Sun. It does this using chromatophores, a special type of cell that contains color; the same cells allow cephalopods like octopuses to produce camouflage. The sea lice can sense light, including the spectral illumination provided by the Moon, just as readily as your eyes can. The lice are known to have two types of internal schedules, one governed by the Sun and one that is apparently linked to the tide. But until recently, scientists were unsure whether the tidal clock was derived from the circadian clock—simply by cutting it in half, for instance, with a tidal timekeeper that runs every 12.4 hours. But it turns out the lice have a distinct tidal clock, which runs separately from any Sun-derived rhythm. It is far more complex than simply slicing the day in two.

Scientists in the United Kingdom collected lice from their home beach on a Welsh island, and measured their activity in a tank where they were exposed to either constant darkness or constant light. The researchers then focused on knocking out or suppressing genes known to play a role in the circadian rhythm. The creatures still swam every 12.4 hours, for several days in a row. Suppressing the circadian rhythm did not shut off the lunar rhythm, showing it is an independent system. In the next few years, marine biologists found similar molecular tidal clocks in animals like oysters, curly crustaceans called comma shrimp, and a mangrove cricket.

Using new sequencing techniques, biologists like Kristin Tessmar-Raible are beginning to understand how animals are pulling this off. Tessmar-Raible studies a marine bristle worm, *Platynereis dumerilii,* which might have one of the most advanced lunar clocks studied so far. The worms follow the Sun for their feeding rhythms, emerging at night to eat. But their spawning cycle follows the Moon alone. The worms use two methods to modulate this lunar clock. They have light-sensitive neurons in their brains, as well as a set of clock genes, related to the same genes found in vertebrates, including you. But the worms' genetic clock also runs on Moon time.

"When we make an appointment, we don't tell someone just what time it is; we also, hopefully, give them the date. We intermingle two timing systems, and that is basically what these organisms are also doing," Tessmar-Raible told me. "There is no voodoo behind it. We have an inner circadian clock. Why shouldn't animals, or other organisms, also have a calendar system?"

Early in her career, Tessmar-Raible read about previous German biologists who studied these biological rhythms and recalls being shocked that anyone could detect a lunar cadence to an animal's daily life, and moreover, that it could be controlled in a lab. Then she attended a marine ecology conference and mentioned it to some marine biologists who just looked at her, she recalled. "They were like, 'Yes, of course. You don't know about the famous coral example? This happens everywhere.'"

Ancient people knew it, too, though for different reasons. Aristotle knew the sea urchins would swell with the Moon because fishermen had learned that mussels, urchins, and some crustaceans are larger—and worth more money—when their gonads are swollen, ready for reproduction. If you cut open a fresh-caught crab, its reproductive organs will be larger or smaller depending on the Moon's phase.

The animals have their own practical reasons for this calendar keeping. A marine gnat called *Clunio marinus* lives along the Atlantic coast of Europe, where scientists have studied its chronobiology for many years. The gnats mate like other insects, with males fertilizing eggs the females laid previously. Because the gnats must wait for an extremely low tide to keep their eggs safe, they evolved the ability to notice the Moon's phases so they could predict when the tide would flow and ebb. Females will lay their eggs in the lowest levels of the intertidal zone when the tide is at its feeblest, during new Moon or full Moon. Corals do this, too. Levy and his colleagues found that corals have light-sensing neurons that allow them to perceive moonlight on the water. They even have genes that activate in sync with the Moon's cycles of waxing and waning.

THE MOON'S RHYTHMS also play a major role in daily life on land. Animal and plant kingdoms are chockablock with organisms that live by the light of the Moon. Owls become chattier under a full Moon, and some white owls hunt more successfully during that phase, as the Moon-illuminated white plumage of a descending barn owl apparently terrifies its prey into stillness just long enough to be caught.[10]

Antlion larvae, called doodlebugs because of the swirly patterns they leave in the soil, hunt by digging funnel-shaped traps for insect prey. The doodlebugs rebuild their funnels every day, and they are larger around the full Moon and during the new Moon. In experiments that kept antlions in the dark with no natural moonlight to guide them, their trap building still followed the lunar cycle for a while before it was lost, suggesting the creatures need their cycles set by the Moon's light.

The same may even be true for another kingdom of life: plants.

A Scottish scientist named Peter Barlow spent much of his career studying leaf movement patterns in greenhouses, on the International Space Station, in total darkness, and in other unpleasant situations. He studied the movements of the stems of peppermint seedlings and found that the leaves are significantly affected by the Moon's gravitational force, not just its light. The plants know that the Moon cycles above.

In 2008, Barlow showed for the first time that leaf movements of baby bean sprouts were correlated to the oscillating lunar tide.[11] In the morning, say, the bean sprouts' leaves might point straight up; at night, they may lie horizontally. Barlow showed a relationship between bean sprout leaf movements, the 24-hour cycle between night and day, and a 6.2-hour tidal cycle. This is no black magic, as Barlow himself wrote; rather, the plants and the Moon must be in some form of continual communion, with the plants responding to the rate of gravitational change as the Moon swings around the planet.

After that discovery, Barlow dedicated himself to studying the multitude of ways in which the Moon affects plant life. He found a tide-related variation in the rate at which plants elongate their roots, correlating to about 24.8 hours—two full tidal cycles. He found a daily variation in tree-stem diameter and in electrical potential. The latter is related to a plant's ability to make food from sunlight, and Barlow showed that the variation correlates to the lunar tide—suggesting a tidal force at work within the wood of trees.

Barlow ultimately extended his experiments to space. In research published in 2015, Barlow and his colleagues found seedlings of thale cress, science's model plant organism, responded to lunar gravity's tidal forces even while growing on the International Space Station.[12] The ISS orbits Earth about every 90 minutes, which would correspond to feeling two high and low Moon tides every hour and a half. In one experiment, astronauts grew the plants in normal light, and found their leaf movements changed every 45 and 90 minutes. In a second experimental setup, the astronauts kept the plants in darkness, attempting to eliminate any light-related circadian signal. The plants still moved every 45, 90, and 135 minutes—like clockwork. They followed the rhythm of the Moon's tide even while they grew off this planet, implying that they can sense the Moon's presence. What this means for growing plants on the Moon itself, perhaps in future settlements, is less clear.

LUMBERING, SHOULDERED FISH; moonstruck corals; speckled sea lice; even peppermint leaves have evolved to live under the influence of the ever-changing Moon, their evolution shaped by the swell of the tides and the light in the sky. But the sublunary influence does not end with them. The presence of a circalunar clock across so many species shows that their clocks probably evolved *before* walking fish, before the conquest of land by both animals and plants. Moon watching may be one of the oldest tools in evolution's kit.

◯ •

IN OUR WALK through time, we have observed the Theia impact, the resulting synestia, Earth's Hadean eon, and a sunny afternoon in the Devonian. Now fast-forward to a Tuesday four hundred million years after that. We are still near Earth's equator. It's sunny and warm, and the glaciers have begun to retreat. It is three hundred thousand years in our past.

By this date, *Homo sapiens,* a primate that favored walking on two of its available limbs, had emerged from the Pleistocene grasslands of equatorial Africa. This newly evolved, highly capable animal lived in groups and worked together, and made note of the Moon's rhythms.

Like our fish relatives before us, ancient humans knew the tide would change the appearance of shorelines throughout the day. They knew this would help them hunt, and they learned their quarry's habits long before Aristotle ever wrote a word about a sea urchin. Early human beings figured out that the Moon and its tides would help them travel across seas to distant shores.

A few artifacts remain from the oldest known human settlements, scattered throughout Africa, Indonesia, and Europe. In many of the totems and talismans, early humans apparently began to bring the Moon down to Earth, etching it in bone and stone and painting its phases on the walls of decorated caves.

During the Apollo program in the early 1960s, a journalist named Alexander Marshack went looking for the earliest representations of the Moon, trying to understand how human culture had culminated in the Moon shot. He dug through dusty filing cabinets in stale archaeological archives that were filled with bones and rocks etched by prehistoric people. The task was monumental and he could not figure out how to begin, so one day, sifting through collections at the French Museum of National Antiquities in Saint-Germain-en-Laye, Marshack decided to limit himself to just one cabinet.

He walked to the first cabinet in the main exhibit room and opened the drawer. The cabinet's upper left corner began in the Aurignacian period, representing a culture that arose in Europe some thirty-two thousand years ago, and the lower right corner culminated in the Magdalenian period, ending around twelve thousand years ago. "It was a fair sampling, then, of man's markings across nearly 20,000 years," Marshack wrote in his book *The Roots of Civilization*. He started in the upper left corner, gently lifting bone shards and rocks. He soon palmed a small, smooth ovoid segment of antler. It had been found in the Dordogne region of France, at a prehistoric settlement called Abri Blanchard, in 1912.

Marshack had picked up a jeweler's loupe for $1.50, and he donned it now, squinting at the bone. Its main face was pocked in a pattern of pits, some of them arced like a comma or an apostrophe and some of them circular. The pits traced a long and winding pattern around the surface, like a meandering stream or a coiled snake. Marshack turned the bone shard in his hand. He realized some of the commas arced from right to left, others from left to right, and realized they were sequential. He saw the phases of the Moon.

"I had lived with the waxing and waning Moon in thought so long, had been watching the phases in the sky," he wrote, "that I now felt at home with this odd, serpentine figure."[13]

He assumed the bone was a form of lunar notation, beginning around the new Moon, when the first crescent would reappear. To Marshack, its purpose was unmistakable: It was an easy-to-carry calendar, a way of keeping time on the road as hunters followed herds of bison, mammoths, or deer. It was carved thirty-two thousand years ago.

The Blanchard bone remains controversial, in part because Marshack, as he readily admitted, made so many assumptions in his analysis. What's more, the comma marks are imperfect owing to the rough tools of the time and to the likelihood that the bone carver missed a few Moons during cloudy nights. But Marshack went on to find scores of other so-called lunar calendars and to publish several volumes of academic papers and books on the astronomical thinking of early humans.

For Marshack, the idea of a lunar record-keeping system was more than an anthropological marvel. It became the basis for a much broader argument about the cognitive abilities of prehistoric humans and their hominid kin. Humans of the Ice Age were much more advanced than scientists had thought, he argued. Comma-shaped bone etchings and cave wall stonework not only represented the Moon and its cycles but also showed that the Moon had enabled ancient humanity to understand time. Keeping track of time's passage allowed people to connect the cycles of the natural world to the cycles of their lives.

Cave art found throughout the Dordogne region of France punctuates this point. The Venus of Laussel, a carving in relief, is thought to represent a connection between human fertility and the lunar cycle. The figure, carved around twenty-seven thousand years ago, depicts a voluptuous woman with long hair. Her left hand rests on her abdomen and her right arm is raised, clutching a bison horn that has been etched with thirteen notches. Her face appears to be turned toward the horn. The horn could represent the number of sidereal months, or lunar cycles, in a solar year—which happens to be approximately the same number of menstrual cycles in a solar year.

The rise and fall of chemicals in a human woman's body that dictate whether she will become pregnant is still poorly understood, especially relative to other biochemical processes. This is despite a half century of research by the fertility-industrial complex. The Moon may have guided this cycle, especially long, long ago, when our apelike ancestors began living in communal groups and following game. What the Moon does to us today, in an era of pervasive and pernicious artificial light at night, is less clear. But even the most up-to-date evidence suggests the process of making new humans is not divorced from the Moon's power.

To START, the average menstrual cycle, meaning the time between the first days of two consecutive menstrual periods, is 28 days long.* The length between consecutive appearances of the full Moon is 29.5 days. And this synchronicity is not even limited to us. Orangutans and gorillas also follow a 28-to-30-day cycle in which the production of estrogen, variations in body temperature, and the timing of ovulation change according to the same cycle as the Moon.

Though the mechanism is not yet understood, this connection is probably not a coincidence. Imagine a clan of primitive *Homo sapiens* living in caves and rudimentary shelters on the savanna. The females, especially those nursing their babies, would stay close to home and forage for nuts and berries nearby. The males would leave for days at a time to hunt. During the full Moon, the open country is bright for many hours after the Sun sets—a good time to travel, or to hunt nocturnal prey. "Our prehistoric ancestors needed all the light-time they could get," Anthony Aveni, professor emeritus of anthropology and astronomy at Colgate University and one of the founders of the field of archaeoastronomy, told me.

During the new Moon, on the other hand, nights are prohibitively dark, so the men may have stayed home and, maybe, in bed with the women. Women who were ovulating during the new Moon phase would

---

* There is no *typical* menstrual cycle; anything from 21 to 35 days in length is considered healthy. But peer-reviewed research shows that 28 to 29 days is the statistical average.

therefore be more likely to become pregnant. Over evolutionary time, our bodily pacemakers may have indeed synchronized with the Moon's cycle, but modern research has only just begun to clarify how.

IN 2020, CHARLOTTE HELFRICH-FÖRSTER and her colleagues at the Julius-Maximilians-Universität Würzburg in Germany examined long-term records from twenty-two women who tracked their menstrual cycles over time, up to thirty-two years. She told me she expected to find a weak correlation with the Moon's cycle, if any, but she found the opposite. The lunar cycle strongly influenced the onset of menses, with several women beginning their monthly periods at the time of the full Moon. This means that for some women, their most fertile times, near ovulation, would be two weeks later, when the Moon is new and dark. There was also a peak during the new Moon.

Two participants often reported brief menstrual cycles several days shy of the twenty-nine-day lunar period. But shortly before they became pregnant, the women's cycles lengthened to match the Moon's.[14] Helfrich-Förster said she cannot yet explain why, but it seems possible that synchrony with the synodic month—the time it takes the Moon to return to the same alignment with Earth and the Sun—may have conferred some benefit to their fertility. She believes there may also be a gravitational reason, because several women's cycles aligned more strongly with the Moon when it was closer to Earth.

The study has the usual caveats, including the fact that it included only a couple dozen people. But poring over Helfrich-Förster's data sets, I found the lunar signal to be unmistakable.

"When you think about all the factors that are involved in this, the fact that you can show a signal at all means there's something there," Katherine Sharkey, a sleep scientist at Brown University, told me. "Why wouldn't we be susceptible to these rhythms? It's super basic priming: If you are an amoeba, don't reproduce at noon, because you will get dried up. You have to be able to organize your reproductive behavior."

Cave artwork dating to twenty thousand years ago suggests people at that time may have even figured out the link between the lunar cycles and fertility. They could have used the Moon and stars to plan their pregnancies, according to analysis by Michael Rappenglück, a science historian and astronomer in Gilching, Germany.[15] This would be a more intentional sort of fertility awareness than the pure biology Helfrich-Förster observed. The rising and setting of stars at dawn, called the heliacal rise, has been used since prehistory to mark seasons. Combined with the Moon's cycles, star watching could be a potent form of family planning, according to Rappenglück. A woman might keep track of her menstruation in order to recognize the full Moon cycle closest to the dawn rising of the bright star Betelgeuse, marking Orion's shoulder. If she conceived after that period, she would deliver her baby when the star was setting at dawn. This meant her birth would happen in spring, after the harsh winter, when food was more plentiful for the nursing mother and her infant.

We can't go back to the Magdalenian era to ask mothers whether they used this form of family planning. But we know for certain that this deliberate seasonal synchronicity is true for many animals, including the creatures that undertake Earth's most stupendous migration.

Each year on the Serengeti, an ancient grassland that borders Tanzania and Kenya, a horde of some 1.7 million wildebeest move clockwise from north to south and back again. The herds cover more than five hundred miles one way in search of rain-fed grass, and nearly 250,000 of them perish on the journey. The animals give birth to their calves, nearly 500,000 new wildebeest, while in the southern part of the Serengeti during austral summer in a narrow, three-week span in January and February. Calves born outside this window almost never survive—they fall victim to predators, like crocodiles and lions, that follow the herd through its great migration. Calves born at the right time will be old enough, and strong enough, to make the hard journey at just three months of age.

But for baby wildebeest to be born en masse, their mothers must also mate en masse, ensuring they conceive their young at the same time. Even though scientists have struggled to pin down an exact window for these dates, the estimated conceptions apparently always fall between April and May, in a window of time determined by two consecutive full Moons, immediately following the southern fall equinox.

It's probably worth noting that near the equator, where the Serengeti lies, the waxing and waning Moon is a powerful luminary. It seems possible that wildebeest, just like humans, synchronize their lives with a reliable clock in the sky.

○ ˙

THE MOON CAPTAINED the origins and evolution of life, and it probably influenced our early biology. Its regular appearance and disappearance certainly dominated the way we began keeping time. And the Moon most certainly held dominion over the early human psyche. Humans have long assumed that the Moon affects us. The words "lunacy" and "lunatic" derive from it. For millennia, doctors have claimed that the Moon affects biology. Hippocrates himself wrote that "no physician should be entrusted with the treatment of disease who was ignorant of the science of astronomy." In the 1990s, a survey showed that some 81 percent of mental health professionals believed the full Moon changes behavior.[16] Many of the changes people often associate with a full Moon—erratic conduct, increased crime, general "lunacy"—are impossible to prove. But while the epidemiology is hard, there actually are some correlations between Moon phases and human activity, from insomnia to mania. Full Moons are known to reduce sleep in adults and children, which can make some erratic behaviors more likely. Who isn't cranky after a lack of sleep? For reasons no one can quite figure out, hospitalizations for things like cardiovascular problems and some kinds of hemorrhages increase during certain Moon phases. Medical findings are mixed, but studies show aneurysms are significantly more likely to rupture during the new Moon and full Moon phases, when the Moon is aligned with the Sun. The mechanism for this is unclear, especially because changes in the Moon's gravitational pull would be so slight. Yet even a minuscule effect on the thin wall of a brain aneurysm could change the rupture pattern, note the authors of a 2017 study.[17] The Moon may play a role, albeit a tiny one, in the flow of blood through the human body. The litany of anecdotes is broader than the peer-reviewed studies on the subject. Go into any hospital around the time of the full Moon and you will notice the correlation.

Beyond affecting physical health, the Moon has observable effects on mental health, too.

Psychology has come a long way since the days of so-called lunatic asylums. Most scientists scoff at the notion that a full Moon would make anyone act differently or engage in the risky behaviors associated with full Moon cycles. But there is some new evidence that there is in fact a relationship between the phase of the Moon and human behavior. In 2018, German scientists showed that in some patients with bipolar disorder, switches from depression to mania coincided with the synodic month, the 29.5-day recurrence of the new Moon.[18] In one patient, manic episodes surged because the patient's wake and sleep cycles synchronized with the tidal rhythm instead of the circadian rhythm.

Disrupted sleep might be to blame for this. The lunar cycle is also known to affect sleep, through both light and an apparent gravitational effect that scientists are still unsure how to explain. In a major 2013 study, researchers studied volunteers in a sleep lab where they were isolated from artificial light at night. The volunteers were not told that the study would be looking for any Moon-related effects. The researchers found that around the full Moon, deep-sleep-related brain activity diminished by 30 percent, total sleep dropped by twenty minutes, and it took people an average of five minutes longer to fall asleep in the first place. The study was the first reliable lab evidence that the Moon can modulate sleep in humans, even apart from its light.[19]

The Moon has many subconscious effects on humans, in early prehistory and now. But its pull on the human consciousness is arguably even more profound.

Archaeoastronomers say the Moon played a potent role in the increasing complexity of early human thinking. In the Paleolithic and into the Bronze Age, lunar symbolism likely enabled humans to understand, or at least to relate to, the otherwise mysterious concepts of becoming, birth, vanishing, death, resurrection, renewal, and eternity.[20] While the silvery crescent of a new Moon is emblematic of possibility and fertility, the amber-toned sickle shape of the waning early-morning Moon can represent endings, or an easing into nothingness.

Ancient people followed the Sun and the seasons, but they also noticed the phases and the longer motions of the Moon. Its utility as a daily, monthly, and even seasonal timekeeper was unequaled.

"If you want to keep time by the Sun for anything longer than a day, you're going to start running out of fingers," said Aveni, the archaeoastronomer. "But if I were to tell you, 'We'll meet at the next quarter Moon,' that's an easy way to mark a couple of weeks. We can say when we see the first crescent through the west, we can say when the Moon disappears, we can say a full Moon, and not be off by more than a day either way."

Archaeologists debate whether ancient carvings and images of the night sky were created for timekeeping purposes, or for art's sake, or for more advanced ritualistic uses we can only guess at. But what's clear is that the motifs recur over time and across great distances.

Many Paleolithic people in Europe and the Mediterranean venerated (or at least frequently painted) bulls, whose horns represented the Moon in later cultures throughout the world. The paintings of Lascaux Cave, which date to seventeen thousand years ago, are best known for their evocative depictions of both horses and bulls. Curvy animals cover the walls and the ceiling of the cave, along with dots that might represent stars and the Moon. Beyond the cave entrance, past an area called the Hall of Bulls, a visitor enters a dead-end passage known as the Axial Gallery. Red aurochs, an extinct form of cattle, stand in a group, heads together. A huge black bull stands opposite them. Across the gallery, a pregnant horse gallops above a row of twenty-six black dots. The mare is running toward a massive stag, whose front legs are obscured behind thirteen additional evenly spaced dots. Dots and dashes accompany animals throughout Lascaux and other caves, from Chauvet in France to Altamira in Spain.

Some anthropologists think the animals and their accompanying notations may be intended to represent seasons. In Europe, bovines calve in the spring, and horses both foal and mate in late spring. The deer rut takes place in early autumn, and ibex, a type of wild goat, mate around the winter solstice. Some archaeologists believe that in Lascaux, the thirteen dots depict the full Moons of the lunar cycle, and the twenty-six dots may represent a sidereal month, roughly the time it takes the Moon to orbit Earth with respect to the stars.

Taken together, the animals and the dots may represent the first steps toward a primitive calendaring system, perhaps to convey seasonal information about game.[21]

But these early suggestions of lunar timekeeping still lack an explicit representation of the Moon. They are just the first steps along the path from a biological system of timekeeping toward a civil one. It was only after the Ice Age in the Pleistocene that early people would come to use the Moon as not only a marker of time but a forecaster. To do this they would need something more human, and maybe more mystical, than the lunar rhythms pulsating in their cells. They would need more than a carved bone shard etched with commas.

We had come a long way from our ambling piscine ancestor stranded by the low tide. Now we needed tools to hunt those fishy cousins to sustain our growing communities. We needed a way to govern our days, allowing us to divide and conquer time and truly take over the world. Reckoning time by weeks and months rather than days gave our ancestors the ability to plan far ahead. With a reliable, natural calendar, people could agree upon dates for harvests, hunts, ceremonies, feasts, slaughters, and wars. They could create the first societies and the first literate civilizations. To do all of this, they would use the Moon as their guide.

CHAPTER FIVE

The Beginning of Time

HILARY MURRAY CROUCHED over her kitchen table and took a green highlighter to the map she'd spread between us. She traced the roads we'd traveled that day, then marked the ancient stone circles of greatest merit. She noted the site of Crathes Castle with a pencil. The estate, built in the sixteenth century on land donated by Robert the Bruce in 1323, apparently didn't merit a mention on the map. But to Scottish archaeologists Hilary and Charles Murray, the castle's grounds contain one of the most important historical sites in the country. Depending on the way one interprets it, it may well be one of the most important sites in the world.

BESIDE THE TAN castle building, beyond the ornately sculpted hedgerows and formal gardens, there is a simple grass field edged with willow and alder trees. Beneath the field are twelve sunken pits, carved ten millennia ago by Stone Age hunter-gatherers. They were hidden by time and grass until the parched summer of 1976, when a dry spell liberated them from the soil, allowing archaeologist Gordon Maxwell to spot them from a small airplane.

Maxwell and others were conducting a survey for the Royal Commission on the Ancient and Historical Monuments of Scotland, aiming to see what emerged from the drought-stricken land of Aberdeenshire. Maxwell noticed a weird wrinkle in the valley of the River Dee, in a glacier-carved field near Crathes Castle. He snapped some black-and-white photos.

The developed images showed the clearing, called Warren Field, etched by modern-day plow marks. But beneath them was an unmissable set of enigmatic shapes: two thick rectangular shadows that appeared to be buried ridges, and, more oddly, an alignment of pits stretching from southwest to northeast. Maxwell and his colleagues published their findings in a journal called *Aerial Archaeology*.[1] The ridges under the ripening crops were interpreted as the remnants of a large timber hall, interesting but otherwise unremarkable, especially in a country littered with the ancient leavings of Neolithic people. The alignment of pits was unexplained.

THE SAME YEAR, Charlie Murray started work as a full-time archaeologist for the city of Aberdeen, with Hilary joining the department four years later. During the next few decades, the pair worked to document Middle and New Stone Age relics and ruins in the region, often being called into service when developers announced plans for new buildings and roads. As they drove me through Aberdeenshire in their blue Peugeot, Hilary kept pointing out past excavations. But it was not until 2006 that the duo excavated the unusual alignment of pits at Warren Field.

The Murrays, Irish by birth and bearing, are now semi-retired on their sheep farm and only occasionally take on new excavations in between visits with their grandchildren in Glasgow and Australia. Hilary has a shock of orange hair, a neatly clipped brogue, and a fierce stare, but just beneath the surface is a kind heart and a wellspring of knowledge. Charlie is graying, slowing a bit with arthritis, and as easy to laugh as anyone I've ever met. The wall behind their kitchen table is papered over in family photos: Charlie and a grandson, maybe a year old then, laughing heartily; Hilary smiling with a grandbaby in her arms; a field full of golden flowers and sheep; the family posing on a balcony in Tuscany one year for Charlie's birthday.

Even in their semi-retirement, the Murrays do not flinch from the type of labor their ancestors, and my own, toiled at for centuries. They rise before dawn to check the lambs. They mash the potatoes and then bake them with lots of butter. They walk uphill, into the woods, simply for the sake of walking. Hilary led me through their property, up to the top of a hill amid thickets of willow and alder like the ones at Warren Field. She wanted to throw a stick for her Border collie, Banjo, and to check on a badger den. The Murrays wanted to show me the mouth of the River Dee and its colony of harbor seals, so we drove to a nearby visitors' center and then hiked straight through a thatch of gorse, an evergreen bush that produces brilliant canary-yellow, coconut-scented flowers and is completely covered in spikes. The gorse caught on my coat and hair and Hilary and Charlie both beat me to the beach.

The Murrays are hardworking and hilarious and warm and razor-sharp, and by the end of a long weekend touring Aberdeenshire's gentle hills, I felt like part of their family. Back in the kitchen, Hilary pushed her hair out of her face and peered at the map, showing me the route we had taken through Aberdeen and back in time.

$$\bigcirc \cdot$$

IN 2004, Hilary and Charlie were hired to excavate the buried history of Warren Field so that the authorities at the National Trust for Scotland could determine how to protect the site. Hilary's PhD research focused on Viking-era wooden buildings, so she is used to handling complex excavations involving structures that were used in multiple ways. As the aerial photos suggested they might, the Murrays and their team found the remains of an eighty-foot-long, thirty-foot-wide timber hall dating to the Neolithic, also called the New Stone Age, which corresponds to roughly 3800 B.C.E. The building contains traces of cereal grains, which means it was used during the advent of pastoralism and agriculture. The Murrays carefully excavated the timber hall's remains, including bits of wall posts and roof supports. They collected ashes and burned wood fragments to scrutinize how much carbon-14 the wood contained. The Murrays were able to obtain ages for the samples by comparing the amounts of carbon-14 in the wood fragments; the older something is, the less C-14 there is to detect. In the same way that isotopes of oxygen tell us about the creation of Earth, the Moon, and other rocky worlds, the Murrays' carbon dating procedures revealed the age of the hall and the time when it was used.

Then they turned to the pit alignment nearby. The connection between the two is not clear. But as the Murrays told me, the people who built the timber hall would surely have known about the series of sunken holes carved by their ancestors many centuries before.

Teams of people working together, heaving giant timbers into place, must have had some ideas about the nearby scoops dug out from the earth. They surely knew the field was an important place, though they might not have been able to articulate why. With the passage of ages, the pit alignment's origin story may have been subsumed into founding myths, stories of ancestors, and tales of mythical beings lost to time.

It is almost impossible not to think of these vanished people and their stories everywhere you go in Scotland. It is a country of legends. The national animal mascot is a unicorn. Windswept Aberdeenshire seems full of ghosts. Visitors to Crathes Castle are amazed by the giant, ornately groomed eighteenth-century yew hedges that loom overhead, and they enjoy castle visits, a zip line, and other tourist activities. But amid the manicured gardens and visiting gentry, a low-grade awareness of ancient ways, ancient *beings,* followed me like a shadow. It follows Charlie, too, wherever he and Hilary go. "I think the castle is boring. It's too new for me. I like them older," he said, only half kidding. The castle, completed in the sixteenth century C.E., is three times more distant from the time of the timber hall as Generation X is distant from the time of Christ.

And as the Murrays discovered, the pits were even older, dating to around 8200 B.C.E.—10,200 years ago. The pits were also used far longer than the hall. They are huge, many of them eight feet wide and five feet deep. People dug them using shovels fashioned of deer antlers, animal shoulder blades, and wood, exerting considerable effort. Despite their builders' careful work, the pits would occasionally slump in, and soil would cover deposits of ash and pollen. After some time, people would excavate the pits once again, until more spoil tumbled into them. This pattern continued for two millennia.

This is a remarkable span of time. Very few structures have lasted so long—consider the Great Wall of China, about 2,300 years old. The Roman Colosseum is about 1,950 years old. Notre-Dame Cathedral, 860 years old, lost its spire in a devastating inferno during the week I visited Scotland. Now imagine a monument lasting more than *ten times* as long as Paris's storied church. The pits of Warren Field are 5,000 years older than even Stonehenge, a much more famous monument to the sky, towering above England's Salisbury Plain five hundred miles south of Aberdeen.

The fact that the pits were dug and re-dug over such a long period of time is evidence of their importance, Hilary told me. She and Charlie are careful scientists, and they hesitated to draw conclusions about the meaning of the pits—that would be left to another landscape archaeologist—but both Murrays agreed that they meant something important, probably to do with astronomy, and almost definitely to do with the passage of time. They were not simply animal traps or mere fire pits.

"It's very strange to create a monument in negative," as Hilary put it to me, after a dinner of lamb steak and Irish-style mashed potatoes. "I'm quite sure they do represent time. But were people realizing they mark time in years, or in generations? Could they have been thinking of time across generations?"

The Murrays and their collaborators published a book in 2009 about their excavation, musing that the rounded pits may have had something to do with astronomy.[2] They considered it a linear monument, meaning they imagined its users would stand near one pit so that the others stretched out before it. They likened the alignment to Stonehenge. That got Vince Gaffney's attention.

GAFFNEY, A PROFESSOR at the University of Bradford in northern England, has made a career of studying earthworks made by ancient humans from the North Sea to Croatia. He speaks carefully in a northern English accent, like an academic Paul McCartney. He is comfortable musing on the quotidian details of a Mesolithic Scottish lifestyle, and equally comfortable making a bold claim: that the concept of time itself, and our place in it, originated in his homeland.

The iconic monument of Stonehenge is exceptionally well studied, but like any unspeakably ancient artifact, it contains lingering mysteries. Near its rock pillars, Stonehenge has a long rectangular ditch called the Cursus, which is older than the rest of the monument. In 2011, construction crews digging out a new parking lot for tour buses found some pits within the Cursus. On the summer solstice, the eastern pit aligns with the rising Sun and the western pit aligns with the setting Sun. Gaffney set out to interpret these pits, and went searching for other Stone Age monuments with similar features. His colleague Simon Fitch, a seismic mapping expert at the University of Bradford, showed him a copy of the Murrays' book.

Gaffney examined the ghostly rectangle of the ancient timber hall. He examined the pits. He studied Hilary and Charlie's hand-drawn renderings of each, which depicted their placement and made them easier to discern than Maxwell's grainy aerial photos. On the printed page, the pits were aligned vertically, the rough circles tracing a jagged line from north to south. Gaffney looked at it, squinted, and craned his neck.

"Simon," Gaffney said. "Turn the book round."

He says he saw it instantly: "It's the phases of the Moon!"

As he recalled it later, Gaffney recognized the pits as a structured monument. To him, the alignment was not linear, the way the Murrays saw it, forming a line from top to bottom. Instead, the pits formed a distinct curve from right to left, bent to the horizon. And it struck him as vital that there were twelve of them. They started out small, gradually growing larger until reaching the largest pit in the middle. Then they grew smaller again. The smallest pits are also the shallowest. Gaffney was captivated. He mustered a team of archaeologists for a second excavation, in 2013. They brought ground-penetrating radar, equipment to study electromagnetic induction, magnetometers, and a suite of other geophysical tools. The team did not find any new features, but they took a fresh look at the pits' sizes and carefully mapped their orientation on the landscape.

The alignment made sense to Gaffney, but its orientation was unusual; he couldn't figure out why its makers chose this spot, and this arrangement, to carve Moons into Earth. He stood near the center pit and looked up. The spot affords a clear view of a V-shaped pass over the Grampian Mountains, called the Slug Road.

The area was forested in hazel and alder ten millennia ago, but the pass would have been visible when the trees were bare in midwinter. This is the time of the winter solstice, when the Sun reverses its course across the sky to move northward and lengthen the days. Gaffney told me that when he realized this, he felt as though the heavens were shouting at him: This field not only mimics the phases of the Moon, it's an astronomical alignment, too.

Armed with sophisticated landscape maps and three-dimensional computer renderings, Gaffney and his colleagues decided to turn back the clock to determine whether the winter solstice was indeed related to the Warren Field alignment. They had to rewrite astronomical modeling software, which only cycles backward until about 4000 B.C.E. That's partly because no one thought humans were paying much attention to the workings of the cosmos prior to that. Roughly 4000 B.C.E., some six thousand years ago, corresponds to the oldest astronomical records in world history, from the first literate societies in Mesopotamia. With updated software, Gaffney's colleagues traced the motions of the Sun back to 8000 B.C.E., ten thousand years ago, when Warren Field would have been in use and prehistoric humans wearing furs and carrying stone tools would have warmed themselves by a communal fire.

Sure enough, anyone standing at the center of Warren Field's pits would have had a direct line of sight to the sunrise on the winter solstice. As dawn pinked the snowy hills, the gathered people of prehistoric Scotland would wait for the life-giving Sun to show itself.

The arrival of the Sun after the year's longest night would be a gift in itself. But the pit alignment calls on the Moon, which makes the pits about something deeper than the changing seasons. The people of ancient Aberdeenshire used it to begin their year. Like the turning of a new calendar page to January 1, the Warren Field pits were a time marker, telling their users when to begin preparing for a new cycle of seasons.

In this part of Scotland, at that time, the seasons were crucial to the hunting and gathering of food. It would be scarce for clans of hunter-gatherers, especially in winter. The solstice would fall two or three Moons before any sign of spring, when deer, otters, and hares would emerge and fatten in the warm and greening hills. But the most important food source was the nearby River Dee, home of the salmon.

In February, two Moons after the winter solstice, the tide carries the fish in toward the shore. Just as their ancestors did uncountable eons ago, the fish move out of the sea, toward the land. The salmon are not here to walk, however; they are here to start their next generation.

The tide-borne fish swim upriver, past the harbor seal colony and the gorse that caught my coat, past the grounds of Crathes Castle, up toward the waters of their birth where they will spawn. During bright moonlit nights, they will swim near the surface to use the night's luminary as a guide. For the people hunting these fish, this Moon was the important Moon, offering the longest nighttime light of the month, the best chance to walk the countryside long after the Sun was down and cover as much ground as they could. The river would be so slick with fish, people would have had more than they could carry. But they needed to plan for this event—maybe to ration their food in the meantime, maybe to fashion a net, or to whittle a spear, perhaps to organize the communal festivities that would happen during the salmon run's peak.

But these people had not invented writing yet, and two Moons is a lot of finger counting to remember. They could etch some hash marks on a stick or a bone, just as their distant ancestors did down on the Continent. But as Alexander Marshack showed, this is more useful for marking days gone by, not days in the future. And a stick or a bone sheath can be lost. It would be better to mark each Moon in a perma-nent fashion, tallying up the Moons gone by so you know exactly which one brings the salmon. This is what the pits represent, Gaffney believes. The twelve pits are a calendrical device, representing the Moons in a solar year.

There is no doubt that hunter-gatherers would have been capable of making observations like these. Though they have long been stereo-typed as primitive louts who "tended to be wandering around the land-scape, scratching their backsides," as Gaffney puts it, Marshack and others showed that early humans were making lunar notations centu-ries before they carved the Warren Field pits. The difference at Warren Field is the permanence of the calendar, and its alignment to the sol-stice, both of which make it truly unique. But Warren Field offers some-thing else, too. The solstice alignment shows that humans had made a giant leap in thinking.

While the Moon's phases are a simple way to create small blocks of time, it's no small feat to wrangle that thirtyish-day cycle into one that matches Earth's seasons. A solar year, meaning the time it takes for the Sun to return to exactly the same spot in the sky, usually has twelve or thirteen lunar cycles. That means a calendar that only counts Moons will soon drift from our tilted planet's seasons, and make you miss the salmon run by weeks or more.

To keep Sun and Moon in step—in order to use the Moon as your calendar, to keep time on the schedule by which you live out your life while remaining in tune with the seasons—you need to add a few extra days to the calendar.

This could be as simple as tacking on a handful of days around the winter solstice. The Twelve Days of Christmas are a holdover from this tradition, but Middle Stone Age clans didn't come up with that. The ancient Scots thought of something else. Early lunar calendars would essentially hit a reset button, starting your Moon year anew after the winter solstice. This way, Moon counting would stay in sync with the cycles of nature.

This is what makes Warren Field special, according to Gaffney. It is an annual corrective, a clock reset that allows people to keep tracking time by the Moon, their reliable nighttime companion, without falling behind the Sun-driven seasonal year. By watching the solstice from Warren Field, prehistoric people could use the pits as a timekeeper.

This is a remarkable feat. It's very practical, but it also represents a monumental shift in thinking. The Moon pits, Gaffney believes, mark the first time humans figured out how to predict future time. It was an evolution in skyward thinking that had started centuries earlier, painted on the walls at Lascaux and Chauvet caves, and pitted in bone and stone talismans found throughout central Europe. Even though the cave paintings are Moon markers, and Marshack's bones are time counters, those artifacts were likely not produced with any consistency. There was no universally recognized beginning, end, or sequence of any kind. They had no continuity, and there was no way to connect them to one another, or to the seasons. The pits are different. They can do all of this. They are lunar time *reckoners*. They orient the user in time.

Archaeologists have long argued that an appreciation of the passage of time should not be equated with a system of timekeeping or calendaring. But time orientation, even if it's not through a calendar, is a cognitive leap beyond the simple acknowledgment that day follows night.

It's hard to overstate the significance of this shift in perspective. Today we live by time, which we never seem to have in sufficient supply. Never mind the cycles of the Moon; forget even the rhythm of your own beating heart. Today we count time so carefully, we measure the second as 9,192,631,770 oscillations of a cesium atom. The sexagesimal system of sixty seconds in a minute, sixty minutes in an hour, twenty-four hours in a day, bequeathed to us by the Babylonians, is too inexact for today's world. Financial markets operate in fractions of a second, meaning fortunes rise and fall according to the beat of this randomly chosen quantum phenomenon. Time is the backbone of society. We live in time. But it is a quintessentially human construct. As far as we know, only humans can situate ourselves in time, and mentally move around within it, from the past to the present and to the future. We can only speculate about when this cognitive ability arose in our hominid ancestors, but if Gaffney's interpretation is correct, Warren Field is the first site we've discovered that demonstrates this kind of reckoning. If this is true, humans began to orient themselves in time far earlier than archaeologists have long thought.

$$\bigcirc\,^{\bullet}$$

THE MURRAYS AGREE, as previously mentioned, that the pits were probably used for timekeeping of some kind, and for ancient intentions beyond their understanding. While they stop short of saying the Warren Field pits compose the first lunar calendar, they say Gaffney isn't necessarily wrong.

"We had looked at it as a linear monument. He was the first person who thought of it standing crosswise," as an arc on the horizon as opposed to extending in a straight line, Hilary said.

At midwinter, the solstice sun would have risen above the Grampian range, visible from the central pit when the alder and hazel trees were bare. But the alignment was meaningful at midsummer, too. Around the summer solstice, the Moon would have risen on the southern horizon and remained low, owing to northern Scotland's extreme latitude. It would appear to roll along the tops of the hills before disappearing. People may have used the pits to mark the passage of time as the seasons cycled forward. Maybe each pit represented a phase of the Moon, or maybe each pit represented a lunation (a complete lunar cycle; i.e., a month). Or maybe both were true.

A clan of hunter-gatherers might have sent an emissary to the great gathering place around the solstice time. Maybe they would meet new friends or see old relations there. Maybe they would conduct rituals or ceremonies, whose meanings are now hidden by the veil of time. The Northern Hemisphere's midwinter solstice, the day with the shortest sunlight of the year, has been a time for celebration for thousands of years, after all. It's one of the reasons so many faith traditions hold festivals around the winter solstice, including Christmas.

"The fact that you have this massive wash of people going up and down the river seasonally, it would be quite possible for them to have gathered here at times during the year," Hilary mused.

While we can only wonder what people were doing at the pit alignment, there is one thing we know for sure. After more than two thousand years of continuous use, the pits fell into disrepair. The holes were later filled in, a mysterious choice that spooks Charlie. "That's the truly scary thing to me," he told me. "Were they trying to undo something?"

And around the same time, the timber hall was deliberately burned down. Archaeologists generally agree on the widespread intentional destruction of Neolithic structures throughout Europe. Hilary and Charlie studied charcoal patterns and wood samples from the pits to show that the Warren Field timber hall must have been set on fire deliberately. Oak, which was used to build the hall, would have been difficult to ignite directly, suggesting that other, more flammable material was set ablaze first.

The exact dates are hard to identify, in part because Warren Field was carved by glaciers. For an archaeologist, erosion is a harsh and tricky foe. But the Murrays have one idea. Around 2300 B.C.E., a volcano called Hekla, in southern Iceland, blew its top. Later eruptions from this volcano are known to have cooled global temperatures for two decades. One such event is recorded in oak tree rings preserved in Irish bogs.[3] Under an ash-choked sky, trees barely grew for eighteen years.[4] Researchers have suggested conditions were akin to a nuclear winter, making agricultural life all but impossible in northern Scotland.[5] Perhaps people viewed the cloudy sky as a bad omen. Maybe they thought they were being punished for their skywatching. Or maybe they could no longer see the Moon behind the haze, and thought it had gone, and so they abandoned their earthen monument. This theme of abandonment and destruction recurs in other Moon monuments throughout Scotland, and throughout much of ancient Europe.

TODAY, THE PITS and the timber hall are invisible. Their exact location is known only to the Murrays, Gaffney, and a few others. People visiting the exquisite walled garden of Crathes Castle have no idea what they are missing as they drive up the hill. The pits and the timber hall's remains were carefully filled in and fenced off, to keep curious tourists, rabbits, and other visitors away from the ancient monument and to protect anyone from falling in.

"It is the only time I cried when a site was being backfilled. It just haunted me. I went away and walked in the wood. I just couldn't get over it," Hilary told me.

She visited in 2018 and took three acorns from the site. Now three young oaks from Warren Field grow on the Murrays' property, descended from the very towering trees that probably overlooked the pits ten millennia ago.

Hilary and Charlie took me to see the field, warning me that I would only be able to imagine what lay beneath. Charlie parked the Peugeot along the road leading to the castle and we hopped over a tall, newly installed wood-and-mesh fence. Hundreds of years of plowing had turned the once-rolling glacier field into a flat plain, flanked by woodland. I walked along the perimeter, where the pits are found, and stopped when Hilary told me. I looked down. The pasture was almost comically green, the same bright kelly tone as Hilary's parka. Trees surrounded the field on every side, and I could not see the River Dee just a few feet away, but I could see the Grampian Mountains. I closed my eyes and imagined what this would have felt like ten thousand years ago.

The wind would be the same, for sure. Scottish spring wind bites like the gorse. The trees would look mostly the same. Through their burgeoning leaves, I imagined two hunters approaching, carrying a post that has been strung up with fat salmon, fresh from the Dee. A couple months into the great salmon run, there is still plenty to catch. The hunters are wearing animal skins as clothing. A small party of fishers and hunters trails behind, while ahead of them, still more people prepare fires. There is heather growing among the rocks, glacial erratics that are now being used as standing stones or seats near the outdoor hearth. The cooking fires flank the curved alignment of pits. Inside four of the pits, more fires glow. They are burning atop torches carved of felled timbers. It is early in the fourth Moon of the year.

○ ˙

THE WARREN FIELD pits may have been the earliest lunar monument in Scotland, and maybe the world, but they were hardly the last. Long after the destruction of the Warren Field timber hall, long after the final sad slumping of the carved lunar pits, the people of northeastern Scotland remained in communion with the Moon. In the centuries before recorded civilization, they erected monument after monument to it. These Moon-watching memorials are called recumbent stone circles, and they are scattered throughout Aberdeenshire.

While Stonehenge is by far the most famous stone circle of all, similar arrangements of stones and wood are found throughout northern Europe, from Ireland to Scandinavia and into Germany and Spain, and even parts of North America. Some of them are made from glacial erratics. But most stone circles use quarried rock, a testament to prehistoric humans' aptitude at moving extremely heavy objects across great distances.

Recumbent stone circles, however, are unique. They are found only in northeastern Scotland. Archaeologists have identified as many as one hundred, from crumbling remnants to towering megaliths scattered through the Grampian Mountains, the gateway to the Scottish Highlands. Their name derives from the arrangement of the largest stone, which was laid in a horizontal, or recumbent, position.

Stone artwork is common throughout the British Isles, on both sides of the Irish Sea. Celts, Picts, Viking invaders, and earlier residents of the countries now known as Ireland and Scotland all used granite for artwork, navigation, and messages we can today only guess at. The Murrays took me to several of these rock billboards throughout Aberdeenshire, like the Maiden Stone, so named because it depicts a mirror and a comb—a sexist trope, Hilary scoffed. But many of the Aberdeen stone monuments are more than megalithic messages. They are directly related to the Moon, and reflect its importance to Stone Age, Bronze Age, and Iron Age cultures.

At a churchyard in Kintore, I ran my palm over an ancient etching of a salmon, leaping over a Pictish symbol for a cauldron. Salmon would have been important to anyone living in this region fourteen hundred years ago, when this stone was carved. And they still are. Salmon fishing is big business in Aberdeen. "If you want to fish here, it will cost you a couple thousand a day. We have some of the richest fishing in the country here," Charlie told me.

On the other side of the salmon carving, on the back side of the stone, is an unmistakable shape, adorned with the knotty loops of early Celtic art: a crescent Moon. The Kintore monument's meaning has been lost to time, but the motifs it depicts are still crystal clear.

AFTER A COFFEE BREAK, we drove to Easter Aquhorthies, a stone circle on a hilltop with a view of the surrounding mountains. It dates to the third millennium B.C.E., not the oldest recumbent stone circle in the neighborhood, but a particularly well-preserved one. Its name might reflect the reason why: Easter Aquhorthies may be derived from the Celtic words for "field of prayer." People might have appreciated the stone circle's spiritual significance long after the original rituals for which it was built faded into prehistory.

The circle is a few moments' walk up a hill from a modest parking lot. I was not sure what to expect, though the Murrays had told me about the colors of the rocks and the circle's unusual acoustics. At the top, I was completely caught off guard. Entering the circle was like stepping through a portal in time.

I climbed three stone steps to enter the circle, brushing past a shiny pillar of red jasper. It was one of eleven erect stones forming a circle more than sixty feet across. The circle is completed by the sideways recumbent, which is flanked by a pair of slabs on each side. Twelve major stones in all, for twelve lunar phases in a solar year. The recumbent stone, the largest one, is twelve feet long, pure reddish granite, and probably weighs nine tons. It likely came from the mountains many miles away.*

---

* I marveled at how people could have moved them, but the Murrays said heaving a stone like the recumbent is easy enough, with the right type of rollers. They have done it themselves, during reconstructions.

I stood in the center and listened to the way the rocks amplified the sounds of our voices and the ever-present cutting wind. The circle's acoustic properties remind Hilary of an old cathedral, where echoes travel to surprising places, and moreover, where you *feel* different because of the environment. The sounds and smells of an old Catholic church are designed to be familiar to anyone who has previously set foot in one, on any continent. The stone circles may have played a similar role. It was intimidating. I found myself hovering close to a pillar at a time, because standing in the middle of the circle was unnerving. I felt like I did not quite belong there, the way I have sometimes felt while standing on an altar. Charlie and Hilary stood apart from me and the rock arrangement's acoustic properties transported their voices through the air like some kind of magic trick.

The circles were probably ritual centers where people would gather during certain times of year. Farmers scattered loosely throughout Aberdeenshire were the builders, but the circles are not near any former settlement, so, Hilary told me, they were places people made a special trip to reach. Visitors would have easily been able to see other stone circles from Easter Aquhorthies, and vice versa. It's possible that people used the circles, and probably bonfires within them, to communicate with neighboring clans—friendly or otherwise—using smoke and colored flame.

It's not obvious in the middle of the morning, but the stone circle is special for more than its size. The orientation of the ancient stones to the Moon's cycles is extraordinary. It shows that after ancient people started tracking the Moon's cycles to calibrate the course of their year, the descendants of the Warren Field calendar makers may have also recognized two strange and useful characteristics of the Moon: the lunar standstill and the Metonic cycle. To understand these phenomena, we have to talk about the movements of the heavenly bodies.[6]

If you pay attention to the sky for a few months, you'll begin to notice a double Dutch phenomenon between the heavenly lights. In the summer, the Sun is high overhead, and in winter it sinks low on the southern horizon, reaching its lowest point at the winter solstice. The Moon is a mirror image. In the winter, the Moon is high overhead, where it can give the luster of midday to objects on snow-covered ground. In the summer, the Moon is lower on the horizon, maybe even appearing in your bedroom window.

Remember that as Earth orbits the Sun, it is tilted on its axis by about 23.4 degrees, which is why we have seasons and why we have the solstices, when the Sun appears to reach its highest and lowest points in the sky. The word "solstice" is derived from the Latin words for "sun" and "stand still," because twice a year, the Sun seems to pause in its daily path before reversing direction. But the Sun is not what moves. The Sun and the other planets seem to arc across Earth's sky, but they actually follow the same plane, the ecliptic; our vantage point is what changes as the tilted Earth orbits our star.

For some reason, maybe owing to its cataclysmic formation, the Moon goes around Earth close to the ecliptic, too, rather than circling our planet near the equator. Therefore, the Moon also experiences something like a solstice. It's technically a "lunistice," but very few people use this term; instead, we call it a lunar standstill. While Earth experiences two solstices a year, the Moon takes much longer to reach its highest and lowest points in the sky: 18.6 years, to be precise.*

LET'S GO BACK to the synodic month, the time required for the Moon to complete one cycle of phases. After twelve of those, the Moon will be almost back where it started. The key word is *almost*. All these moving heavenly bodies, Sun, Moon, Earth, and stars, seem to sail around with respect to one another. The Sun and Moon move against the background of stars, and, further confusing the matter, the stars themselves also seem to move because of the way Earth is circling the Sun. You've experienced this if you've ever gone stargazing at different times of year. At midlatitudes in North America during the fall, one of the most recognizable constellations is Orion the Hunter, whose three-starred belt appears over the eastern horizon in the early evenings. In the spring, Orion is setting in the west by nightfall. To reach the exact same location in the sky, against the same constellations, on the same night of the year,

---

* An ancient Greek named Meton of Athens figured this out many millennia after the ancient people of Aberdeenshire built their lunar megaliths.

the Moon needs more time. This takes nineteen years, or 235 synodic months.

The lunar standstill and the Metonic cycle are not perfectly matched, but they are very close. Especially close for the unwritten astronomical knowledge of people living in 2500 B.C.E. The ancient Scots must have achieved another giant leap in awareness to connect these phenomena, tying two decades of Sun to Moon, without any system of writing that survived the ages.

The standing stones of Easter Aquhorthies are arranged so that in the years of the lunar standstill, just before dawn in the spring, the full Moon sets in the middle of the recumbent stone, between its twin flankers. Maybe ancient farmers huddled in the middle, where I stood to hear the voices of Hilary and Charlie. Maybe ancient people gathered to view the Moon setting. Every eighteen and a half years, people surely watched in awe as the Moon dipped down toward the precise center of the recumbent.

The orientation of the ancient stones to this multidecadal cycle is astounding. Nearly two decades is lengthy in the life of any modern human. What were you doing eighteen and a half years ago? For a person living five thousand years ago, two of these eighteen-year rounds represented late middle age. Constructing a circle whose stones are arranged to embrace the Moon every 18.6 years requires advanced planning indeed. These people had real skill, deep astronomical knowledge, and, it seems, true commitment. The Moon was clearly vital to people of this era. They knew they had to connect their solar year, the year of the seasons, to the year of the Moon—without it, they could not plan their lives.

These motifs recur throughout Scotland. Alexander Thom, who may have coined the phrase "lunar standstill," first explored this 18.6-year cycle at a megalithic site called Callanish, in Scotland's Outer Hebrides, on the western side of the country. Erected around 3000 B.C.E., the stones are arranged in a cruciform pattern, and are situated to highlight the Moon's most extreme rising and setting points. Thom believed people wanted to track lunar standstills and eclipses, though it's also possible Moon watching would have been practical for daily reasons, too; ancient fishers and island-dwelling hunter-gatherers would have been focused on the tides as well as the skies. In 2016, archaeologists from Australia reexamined the stone alignments and concluded that Callanish is a lunar alignment just as much as Warren Field, Easter Aquhorthies, or any other monument to the Moon.

MANY OF THESE stone alignments are forgotten and broken, but some are being recovered and restored even now, thanks to efforts by the Murrays and other archaeologists. One such circle is on the way to the Cairngorm Mountains, away from the sea and toward the purple, orange-dappled landscape of the Scottish Highlands. The circle is next to a rock quarry that was used until the 1920s, and it sits beside a bunker from World War II. I thought it would be somewhat sad-looking, a relic forgotten by time, like the Maiden Stone. I prepared for another hike to see some pretty stones, and I braced myself against the unending breeze.

When we got out of the Peugeot, the wind lashed at my face, the only part of myself not protected by multiple layers of Polartec and rayon. We climbed a hill, both Murrays beating me to the top, and the stone circle of Tomnaverie revealed itself. It took my breath away.

I looked around. The valley and its hills seemed softer than the area where the sea embraces the Dee; the region here has been carved smooth over the ages by both glacier and plow. The grass was golden brown, dappled with green shoots straining toward spring. Gorse was still poking out everywhere, but now I saw purple heather, too, an inhabitant of the higher elevations. A soft rain was visible over the Cairngorms, a few miles to the west. The tableau may as well have been a postcard in a tourist shop with SCOTLAND printed at the top. It was beautiful. It was very easy to imagine myself in the same spot, four thousand springs earlier, witnessing exactly the same scene—sheep, farms, undulating hills, the gorse, the heather, the wind. And the Moon at first quarter, hanging above us.

Tomnaverie originally contained thirteen upright stones, two of which are now lost, which increase in height from the northeast to the southwest. They are arranged symmetrically and they draw your eye toward the northeast quadrant, which is where archaeologists have found copious evidence of burning.

I walked toward the recumbent. These northeastern Scottish stone circles are usually oriented toward the south or southwest. Although some may have been intended to face the winter sun as it hung low in the sky, there is so much variation among the circles, it's almost a certainty that they faced the summer Moon instead. And there may have been more spiritual reasons for their alignment, as British archaeologist Richard Bradley notes.[7] The west, where the Sun and Moon set, may be the direction of the realm of the dead.

Older monuments like Stonehenge in England, Newgrange and Knowth in Ireland, and Warren Field in Aberdeen are oriented to the movement of the Sun at the solstices and equinoxes. But for newer (a relative term) monuments like Tomnaverie, Easter Aquhorthies, and Callanish, the Moon is of prime importance. And the Murrays and Bradley both argue this is evident in their construction. Tomnaverie's standing stones, made of granite and other igneous rocks, contain quartzite inclusions. These are sparkly pearlescent chips of white that reflect light—like from a bonfire, or from moonlight. The Murrays' property boundaries are marked by stone dykes, many of them ancient and some of which also contain quartzite. "There are times you go up there at night and you catch a glint," Hilary told me. The Moon shining in the fields.

TOMNAVERIE IS ONE of a small group of stone circles whose orientation is so far outside the patch of sky where the Sun appears that it must have been dedicated to the other guiding light of the heavens. But the Moon would not have appeared over the center of the recumbent every year. Rather, Tomnaverie might have been oriented to face the Moon every 18.6 years, corresponding to the Metonic cycle just like Easter Aquhorthies.

The view from Tomnaverie was carefully thought out, according to Charlie, who oversaw an excavation and partial restoration in 2003. The recumbent stone and its flankers obscure the immediate foreground and simultaneously highlight a more distant hill and, more important, a segment of the sky. Tomnaverie's vantage point focuses on the summit of Lochnagar, a mountain several miles away, and the northernmost position of the Moon. Had the monument been built farther to the south, the Moon's position would not coincide with the summit of the mountain.

Bradley walked through eighty-six fields in the glacial basin where Tomnaverie rests, known as the Howe of Cromar, and found six concentrations of stone tools dating to the same age as Tomnaverie. The monument sits at the junction between two different environments, according to Bradley. On three sides, the Howe of Cromar is fertile farmland, but on the west—the direction where the Sun and Moon set—the soil is poor. The monument, and its focal point, lay well beyond the area that was farmed and occupied at the time it was built. As with Easter Aquhorthies, archaeologists think Tomnaverie was meant to connect people to realms beyond their own meager, scattered settlements to places greater than themselves, places they could only access in spirit. Though this is only one interpretation of the place, I couldn't help but see it this way myself.

Tomnaverie has one more unusual feature. Its first boulders were installed sometime around 2500 B.C.E., 4,500 years ago, but the recumbent stone was not added until later. It started out with a ring of thirteen pink and white monoliths surrounding an interior rock pile, or cairn. The inner cairn was arranged so that it linked to the outer ring at thirteen points. Sometime later, people added the massive recumbent stone, lying on its side rather than standing tall. The recumbent effectively closes off the ring of thirteen stones. The recumbent is whiter than the rest, maybe reflecting the snows of Lochnagar in the distance, maybe reflecting the Moon high above. Archaeologists like Charlie and Bradley think the building of the recumbent was the last act in constructing these sites. Placing the horizontal stone is like closing a door, or finishing a circle.

Remarkably, the Tomnaverie stone circle was used as recently as the 1600s C.E., when people apparently built funeral pyres within it. I could not help but wonder what those Scots thought of this monument from four thousand years earlier. For that matter, what did people of the first century C.E. think of it? Did they also believe that it was designed to transport people, both the living and the dead, to another spiritual realm?

WHILE TOMNAVERIE IS imposing during a beautiful spring afternoon, its relationship to the Moon means it was most important, and most heavily used, during the night. It would have looked, felt, and sounded completely different in the dark. Moonlight is reflected sunlight, but much weaker, more silvery, more spectral; it blurs the landscape, shortens perception of depth and distance, and "emphasizes the sky quite as much as the ground," as Bradley put it.[8] Moonlight would highlight the texture of the stones and make the quartz inclusions sparkle. At night, sound carries longer distances, and the stone circles have strange acoustic properties. The beating of drums or the sound of chants would fill the air and travel down the hill, through the howe, maybe even to other stone circles nearby. And fires lit within the monuments would totally change their appearance, too. Flames would spread light and shadows that competed with the moonlight, until the monuments themselves seemed to come alive.

To an American like me, it felt odd to live and move among the ruins of people who died two and a half millennia before the dawn of Christianity. The oldest architectural ruins in my region, the American West, date to three thousand years after that. From my own culture of white European immigrants, the oldest thing in my neighborhood is a stone grave marker from 1872 C.E., practically an hour ago in the history of my country, let alone my people on the other side of the sea. By contrast, the stone circles of northeastern Scotland were built in 2500 B.C.E. and used continuously for millennia. In a way, they are still in use. Hilary, Charlie, and I visited them, after all. We walked among them, sat inside them, took in the views surrounding them, listened to one another and to the wind within them.

The stone circles are imbued with a sense of mysticism, but calling them mysterious or enigmatic structures is uncreative. They are monuments with meaning, even if we can't see those meanings anymore. They are incredibly powerful places, erected with purpose and maintained with care.[9] They were built and used for ritual and messaging, maybe to the dead, maybe to neighboring clans. Though we might not know exactly what took place within them, they are undoubtedly places that were planned and used with intention.

The modern world is as removed from the Mesolithic as my house is from Aberdeen. And yet it is not far away at all. Everything has changed, but I could think only of what has not. There is nothing new under the Moon. So many ways of thinking, so many traditions, have remained as constant as the heavens, for thousands and thousands of years.

ON MY LAST evening in Aberdeen, Charlie poured me a gin and tonic—he eschews Scotch whisky—and we sat in the living room. I sipped my drink as he played guitar and sang a blue Irish tune. The Murrays' prized cobalt Rayburn stove was blazing in the kitchen and the fire in the hearth was going strong.

I closed my eyes. It was easy to rewind through ten thousand winter solstices.

In my mind, tall trees shrunk into saplings and asphalt roads melted into wagon ruts, then into footpaths and deer trails. Around me, farmworkers toiled under the eye, and the thumb, of watchful landowners. I sped past Mary, Queen of Scots, through the Middle Ages, breezed by the Picts and the Celts, zoomed past conquering longboats and Roman sentries, kept flipping back, back until I reached the roving bands of hunters and gatherers. I arrived in the Middle Stone Age, on the hills flanking the valley of the River Dee.

Here, prehistoric Charlie and I sit waiting for the sunrise. The Murrays' hill is still covered in willow and alder. Instead of a house with a couch and a stove, there is a small tent made of sheepskin. A wolf dog, taller than Banjo the collie but like him in color and countenance, lounges nearby and occasionally growls at the badgers in their den.

Charlie is sitting on a sheepskin tuft filled with straw. Instead of a guitar he is holding a primitive harp, adapted from his hunting bow. He strums the same sad chords, and the crackling fire plays percussion. The tune we sing and the spirits we share feel warm in my throat. Above us, the fourth Moon of the year shines at first quarter. It is a perfect half circle with its round edge to the west. The sky is filled with stars.

CHAPTER SIX

Early Civilization and the Compass of Time

THE TREASURE HUNTERS had high hopes when they set out into the hills in the summer of 1999. Henry Westphal and Mario Renner knew that the Ziegelroda Forest, in the northeastern German state of Saxony-Anhalt, was full of relics from World War I, World War II, and untold battles over the eons. The forest is also full of barrows, prehistoric burial mounds, often circled by stone cairns, that contain relics from ancient people. The treasure hunters knew digging up those graves was illegal, but they also knew the right kind of Bronze Age loot could fetch a princely sum on the black market. They moved quietly through the forest, swishing metal detectors, looking for an old Nazi battlefield and even older relics of war.

HUMANS AND OUR ancestors have lived and died in this part of central Europe for at least four hundred thousand years. At the end of the last ice age, the retreating glaciers sculpted fertile, rolling plains, which were once covered in grass and inhabited by mammoths, and later filled with forests and deer. The people who hunted those beasts also settled here, and over time, they developed increasingly complex tools and weapons and often buried their dead with them. Ancient tombs across France and Germany spill over with swords and other talismans. Saxony-Anhalt looting laws prohibit anyone from taking them—they are considered property of the state—but this has not stopped amateur treasure hunters from doing it anyway.

After hiking for a while, Westphal and Renner found themselves in a small clearing on a hill about eight hundred feet in elevation. The hill, known locally as the Mittelberg, is about thirty-five miles west of Leipzig, near a town called Nebra. Suddenly, Westphal's metal detector pinged.

The two men started digging with a spade and the upturned earth yielded a trove of objects that had not seen sunlight in 3,400 years. Westphal and Renner dug up two bronze swords, two hatchets, a chisel, spiral bracelets, and a five-pound decorated bronze disk one foot in diameter. In their haste, the looters chipped part of a gold circular inlay on one side of the disk. By nightfall, they had loaded their haul into the car and started celebrating at a pub, newfound riches on their minds.

The damage to the disk was not too severe for a dealer in Cologne, who bought the lot for thirty-one thousand deutsche marks (about fifty-one thousand dollars in 1999). During the next two years, the trove changed hands multiple times on the black market.

In May 2001, Harald Meller had just been appointed director of the State Museum of Prehistory in Halle when a colleague showed him some photos of the Ziegelroda bounty, by now infamous in archaeology circles. Plenty of would-be buyers had been offered a chance at the disk and the related booty, but once people learned of its illegal provenance, interested museums turned it down. Finally, a bar owner who hosted a monthly archaeology-enthusiast drinking club reached out to Meller, claiming to represent the interests of the owner. The seller wanted seven hundred thousand deutsche marks, Meller was informed.

He and a colleague stared at a picture of the bronze disk. The copper had corroded to algae green, but it was inlaid with gold decorations: a full circle, a crescent, a smattering of dots, and two golden arcs. There was also a space where a third arc seemed to have fallen off. To Meller, it clearly portrayed the night sky. Later, he would determine that the stars were the Pleiades, "the Seven Sisters," a collection of bright stars in the Taurus constellation. The crescent was clearly the Moon. The gold disk could be the Sun, but it could also be a full Moon or the red Moon of a lunar eclipse.

The disk would be extraordinary, if it were real. Archaeologists largely considered Bronze Age Europe a wild, warlike backwater, a far cry from the sophisticated, literate societies burgeoning at the time in Egypt and

Mesopotamia. The disk predates the dawn of Greek astronomy by a millennium. If Bronze Age Europeans created something with this much beauty and significance, it would show they were far more astronomically advanced than anyone appreciated.

Whoever made the disk might have used it to study the night sky, maybe to honor it. They might have used it as a tool to track the seasons, and even to synchronize their year. Meller considered it the most significant archaeological find he had ever seen. He wanted a closer look.

An archaeologist focused on Bronze Age artifacts, Meller is a slightly puffy academic in shirtsleeves and rectangular black-rimmed glasses, hardly a swashbuckling Indiana Jones type. But he decided the disk belonged in a museum. He wanted it for *his* museum.

$$\bigcirc\ ^{\bullet}$$

THE STATE MUSEUM OF PREHISTORY in Halle is a ninety-minute train ride from Berlin and a world apart from that wondrous, graffitied, historic city. The state of Saxony-Anhalt, in former East Germany, is the home of composer George Frideric Handel (of *Messiah* fame) and one of Germany's oldest universities, Martin Luther University of Halle-Wittenberg. The city is industrial and increasingly full of migrants, especially from Syria, but in other ways it feels like it has not changed much since Handel walked here. The streets are lined with trees and charming houses. In a butter-yellow Victorian-style home, I noticed hand-crocheted curtains hanging in the windows, like some kind of scene out of *The Sound of Music*.

Today, Halle's most interesting landmark is the State Museum of Prehistory, one of the most well-designed museums in the West and home to one of the most important archaeological collections in Germany.

On the morning I visited, Harald Meller rushed in, breathless and wearing a rumpled shirt, a brown blazer, and a three-day beard. He eagerly shook my hand and offered me a Coke. I was a little early, thanks to the driver Meller had sent to retrieve me at the train station, and he was running behind, the type of researcher who is always distracted by something. He was leaving for Salzburg that afternoon but had promised to show me the museum collections, his carefully curated archaeological finds, and especially the pride of Halle, the Nebra sky disk.

HIS TOUR OF the museum begins with a 370,000-year-old mammoth tibia etched with a repeating series of seven notches. Its meaning is of course a mystery, especially because the notches are less obvious than the comma marks Alexander Marshack identified on the Abri Blanchard bone shards. But it clearly shows the work of a tool user, making deliberate marks in a repeated pattern. Meller believes it is a form of lunar symbolism, an order of magnitude older than the oldest known records identified by Marshack. He believes the Moon has been important in this part of the world for countless generations.

The cognitive abilities of Neanderthals are a subject of heated debate in human evolution research, and Meller staunchly defends them as just as capable as any *Homo sapiens*. I was struck by his care and interest in portraying people of the past as capable, joyful, intelligent beings, a far cry from the depictions of simple hairy brutes I learned about in school. One mural, in a room dedicated to Neanderthals and early humans, shows women laughing as they carry water and nurse their children. They are just as human as you and me. Walking through the museum, you are left with the impression that you've sailed through three hundred millennia only to look in a mirror.

"In most museums, you see these presented as if you are much smarter than these old idiots working with stone tools," he told me. "We always underestimate our ancestors. Always."

In the same room, Meller commissioned a sculpture of a Neanderthal man seated in the same pose as Rodin's *The Thinker*. The young male sits on a ledge, head resting on one hand, peering into the distance, clearly in deep thought.

"A museum is not a book. It is very complex, if you are a scientist like I am, to translate this history. We do this together with artists," Meller explained.

His museum is full of similar finds, presented with great attention to their humanity. The famous sky disk is no exception. But Meller wanted to tell me a story about ancient Germany first.

Down the hall from the Neanderthals rests the bones of a nine-thousand-year-old priestess known as the Shamaness of Bad Dürrenberg. Workers excavating a canal in Saxony-Anhalt discovered her grave in 1934. The woman, between thirty and forty years old, was buried sitting upright, with a six-month-old infant resting between her legs. Their grave was also filled with jewelry; red ochre; the bones of deer, turtles, cranes, and beavers; and mussel shells. Given the care—and the hoard—with which she and the infant were interred, the woman was probably important. Analysis of her skeleton showed that she had two malformed vertebrae in her neck, which could have restricted blood flow in her head and caused involuntary eye movements. She might have appeared to experience trances, or other mysterious behaviors that could have seemed otherworldly; she may have been a powerful shaman, according to Meller. He hired an artist to paint a portrait of the woman as a fierce goddess-like priestess, with an elaborate headdress and face paint.

A visitor to the State Museum walks through the Pleistocene and into the Bronze Age before reaching the sky disk's special room. The portrayals of Stone Age humans were awesome, in the original sense of the word. I was oddly shaken. More than three thousand stone axes were arrayed across a wall like abstract artwork. The room with the mammoth tibia was painted gray, and its walls were rough like an elephant's skin. Meller's vision of past Europeans is all intentional, and it builds to a crescendo with his interpretation of the Nebra sky disk. The entire museum, especially the way he displays its most prized possession, is designed to reflect a culture that was far more sophisticated, for a much longer period, than scholars once thought.

AFTER THREE AND a half millennia of subterranean slumber, the Nebra sky disk is now suspended in a glass case, lit from above and below, giving the impression of the artifact floating in space. The effect even comes through in photos.

It is twelve inches across, roughly the size of a dinner plate. The circle's edges are bent in spots; the biggest dent is an artifact of Westphal and Renner's digging. The entire thing is furred in verdigris, that blue-green patina that covers weathered copper, except for the inlaid sections that are clearly hammered gold. On the right side is a crescent, the size and shape of a few days' old Moon, setting near the western horizon. At the left is a full circle, which Meller believes is not the Sun but a full Moon. Two golden arcs were added to the disk later; one is gone now, but the other adorns the rim, and in antiquity they would have bracketed the crescent and the circle like a pair of parentheses. A third arc, also added later, sits beneath the Moons and stars like a smile. At the center is a smattering of stars that is, unquestionably, the Pleiades.

These stars, known throughout the world as the Seven Sisters, Subaru, and various other names, are one reason the disk is so important. Their appearance in the sky, when cross-referenced with the phases of the Moon, indicates the right dates for sowing and harvesting crops.

The Nebra sky disk is ancient, it's clear. It is rough-hewn, but made with care. It is delicate, yet tough enough to survive the ages. You can see why it has become such a cultural touchstone in Germany, especially in Saxony-Anhalt, the state where it was unearthed. A German astronaut even designed his International Space Station mission patch to incorporate its features. No wonder Meller knew he had to have it. No wonder he took such a risk to get it.

$$\bigcirc \, {}^\bullet$$

A YEAR AFTER he first glimpsed the sky disk, Meller found himself walking into a Hilton Hotel in Basel, Switzerland, en route to meet an amateur black-market antiquities dealer who had offered to sell it to him for a princely sum. A blond woman met Meller at the hotel entrance and invited him downstairs to the restaurant. She led Meller to a thin man with gray hair. After a brief greeting, Meller asked to see the disk so he could verify its authenticity. But the man handed Meller a sword first.

Meller took the sword and examined it. He had handled dozens of artifacts like this, and he could tell it originated around 1600 B.C.E. But Meller knew if he acknowledged its antiquity, the dealer would never part with it, nor the disk it accompanied, without a huge sum of money. Meller lied and told the man it was a forgery. He feigned impatience and asked again to see the disk.

The smuggler did not know it, but across the bar and stationed at every entrance was a cadre of officers from the Swiss Police. Meller worked with them to set up the transaction, and they were ready to make an arrest as soon as Meller laid his hands on the one piece of treasure he really wanted. But Meller was getting nervous. He remembered what one of the Swiss officers had told him: "The most important thing is that you survive." Meller swallowed.

He noticed the man had a small briefcase with him, but he thought the disk was too large to fit inside. The disk was nowhere to be seen. He began to worry he had been set up, and his dread worsened when the man put his hand inside the briefcase, obscuring something within. Meller feared it was a gun. He pantomimed all this for me, sitting across from me in a coffee shop near his museum. He had clearly told this story many times, and relished this detail—fairly enough, because this was probably terrifying, especially for a researcher used to sitting at a desk and musing on the lifestyle of prehistoric humans.

Finally, the man opened his coat. He unbuttoned his shirt. He produced an object wrapped in a towel, which he had draped around his waist. Slowly, he unwrapped the covering, and handed the green copper disk to the archaeologist. Meller gasped: The Pleiades! The golden Moon! He tried not to act too excited. Examining it, he agreed to pay four hundred thousand dollars for the whole lot, including the disk and the bronze artifacts it was found with. He excused himself to go to the men's room, and gave the signal.

The authorities quickly swept in, handcuffing the thin man, the blond woman, and, just for show, Meller. The smuggler was shocked; Meller had assured him Switzerland would offer neutral ground for their illicit transaction. Eventually, the thin man and blond woman would lead Swiss and German prosecutors down the chain of black-market custody, all the way to Westphal and Renner, who reached a plea deal that landed them hefty fines and a short stay in prison—and who led archaeologists to their dig site. But in the meantime, Meller considered his new treasure. He took in the hammered-gold Moons and the precise arrangement of seven stars.

"I was astonished it is so big. It's much heavier than a laptop. It is thick, and heavy, and I was impressed by its beauty," he told me. He immediately believed it was one of the most important finds of the last few decades. "It is so important for the world. It's really important. It is the first picture of heaven," he said.

Archaeologists the world over were astounded, too. Some argued it was a fake, and Meller spent the next several years arranging chemical studies of the metals in the disk, the dirt that covered it when it was found, and fragments of the wooden hilts of the swords it was buried with. Chemical analysis of the metal alloys confirmed that the disk dates to 1800–1600 B.C.E., the height of the European Bronze Age. The copper comes from the eastern Alps, a primary source of copper at the time, and the gold was panned in the Carnon River in Cornwall, on the west coast of England. The tin probably also came from Cornwall. This diversity of metals shows that trade networks at this time were vast and complex.

Five millennia after the pit calendar at Warren Field was dug, Bronze Age humans in Europe had built a flourishing culture, rich with metals and trade, capable of constructing gigantic earthen monuments. In archaeological terms, the people who made the sky disk belonged to the Únětice culture, and their sophistication shows that ancient central Germany was no tribal hinterland. Theirs was a highly stratified society, with ruling princes buried using specialized and ornate rites. Their settlements were dense and covered miles of land. In death, they arranged people on their right sides, with their heads pointing south, so they would be looking toward the rising Sun. Using the celestial bodies, their leaders learned how to grasp time, and how to control its use.

The Moon calendar at Warren Field marks the first time humans learned to orient themselves in time, a major leap in cognition. The Nebra sky disk, a few millennia younger, represents another cognitive leap toward what we might recognize as an early form of modern culture.[1]

Astounding finds from this culture began emerging from the German countryside in the first half of the twentieth century C.E. Archaeological sites suggested a powerful prehistoric culture, its boundaries encompassing much of what is now Europe. The Nazis seized on these findings and positioned the party as the inheritors of this patrimony; in the years after World War II, prehistoric German archaeology slowed down and was limited to a small academic community. Only after the looters Westphal and Renner found the sky disk did interest in the Únětice culture reemerge in the public consciousness.

It would take a few more centuries after the time of the Nebra disk for humanity to use the Moon to create religion and to consolidate power through it, erecting the foundations of modern society. The idea of divinely bestowed power would finally take root in Mesopotamia, the land of Gilgamesh and Enheduanna and Nabonidus, whom we will meet soon. But Meller believes that the Únětice took some of the first steps on this path, guided by the light of the Moon.

In his interpretation, the disk is like a map legend, holding the key that unlocks the sky. To read it, we need to talk about celestial mechanics.

REMEMBER THE DOUBLE-DUTCH jump-rope motions of the Sun and Moon, arcing across the sky during the winter and summer? As the Sun crawls toward its summer height, the Moon sinks on the southern horizon. When the Sun dips low and moves southward, the Moon creeps higher, reaching almost the zenith—the top of the sky—at midwinter. A few times a year, the two cross paths, and that is when we have eclipses. Ancient people around the world, from Egypt to Scandinavia, imagined a great solar boat was responsible for this celestial flip-flopping. According to Meller, the bottom arc on the Nebra sky disk represents this solar boat, or barque.

Together, the golden arcs on the sides of the disk—one of which is now missing, but left a shadow—span an angle of 82.5 degrees. This represents the angle through which the Sun traverses the horizon between the winter and summer solstices in Nebra. This makes the disk a precisely calculated depiction of the path of the Sun, made before math, made before written language, made possible by the Moon.

The shape of the disk's crescent Moon holds the other key.

It is not a sliver, which suggests it's not a new Moon, the first Moon to be seen after a day or so of invisibility. Instead, it is a young waxing Moon, seen after four or five days, hanging, scythe-like, on the western horizon. Why would a time device use a Moon that is at least four or five days old?

The answer was written down eight hundred years later, in Mesopotamia, in the first millennium B.C.E. The Mesopotamians would eventually figure out a more sophisticated calendar reset than the solstice tracker at Warren Field. They realized that to keep the lunar year in sync with the solar year, they should add a thirteenth month every few years. They used the Moon to mark when this should happen.

The oldest compendium of Mesopotamian astronomical knowledge is called the MUL.APIN, which is just the first two words of the treatise. It translates to the name of the constellation "the plow," or the Big Dipper in North America. It names stars and astral phenomena, but more important, it contains two schemes for aligning the lunar calendar with the solar, seasonal year. The catalog lists the heliacal (dawn) rising of certain constellations, the time between the spring and fall equinoxes, and the summer and winter solstices. When the celestial bodies deviate from their predicted positions during these annual events, it's time to insert an intercalary month to fix the discrepancy between the seasons and the calendar.[2] The MUL.APIN writers described an "ideal" calendar of twelve months equaling thirty days each, and used it as a template to figure out when to add the extra month.

For example, the spring equinox of the ideal year takes place on the fifteenth day of the month called Nisannu. This is the first month of the Old Babylonian year, like our modern January 1, but Nisannu was comparable to March or April. If a star or constellation is first visible when expected, then nothing needs to change. But if the star arrives a month later than called for on the ideal calendar template, then that year should be a leap year.

"If the Fish [the constellation Pisces] and the Old Man [the constellation Perseus] become visible on the 15th of Adarru, the year is normal. If the Fish and the Old Man become visible on the 15th of Nisannu, this year is a leap year," the text reads, in one example.[3]

The MUL.APIN's second scheme for calibrating the calendar, and the more relevant for the Nebra sky disk, tracks when the Moon is in conjunction with the Pleiades. On the ideal calendar, this should happen on the first day of Nisannu. The Mesopotamian sky priests note that when this conjunction comes later than expected—indicating that the seasonal, solar year and the human-centered lunar year have slipped out of sync—it is time for a leap year. When the waxing crescent Moon meets the Pleiades in the spring just after the equinox, it is time. A thirteenth month should be added when astrologers observe a young Moon in exactly the orientation that appears on the Nebra disk, which says the same thing, only without words. Is it possible that Bronze Age Europeans figured out this intercalary trick a millennium earlier than the Mesopotamians?

Meller and other scholars can only speculate. No written language from the Únĕtice culture survives, so we will never know how the disk's makers gained such advanced astronomical knowledge. But the disk is evidence that they did. These Bronze Age sky priests would have used the Nebra disk like a clock held against the sky. They would insert an extra month, a leap month, whenever the sky mirrored the gleaming disk in their hands. This would happen every two to three years.

If the sky disk could correlate the solar year with the lunar year, and even calibrate the calendar every two to three years, it would have been an object of great power indeed. The men who make the time are the men with the power. Just as Egypt's pharaohs observed the stars to predict the seasonal flooding of the Nile, the Moons of the Nebra sky disk could have turned its owner into an oracle—and given that person great control over the burgeoning early society of ancient central Germany.

○ ˙

THE NEBRA SKY disk is the most intricately detailed, arguably most beautiful ancient depiction of the heavens. But in the area where it was found, a different group of prehistoric Germans were still at work on the landscape, building earthen calendars of their own.

The Goseck Circle stands in a nondescript wheat field, much like the ancient Warren Field calendar. It was created around 4900 B.C.E., three millennia before the Nebra sky disk, and the discovery of human and bovine skeletons within its boundaries suggests it was used for some sort of sacrificial rituals. Cultural groups rebuilt the circle in the early 2000s C.E., and now you can stand within its wooden stakes, seven thousand years after distant fellow humans did, and watch the sunrise. The Goseck Circle is a solar observatory, not a lunar one. On the winter solstice, the Sun rises and sets between the circle's southwestern gates. The summer solstice Sun rises in the gaps between the wooden stakes.

The proximity of Goseck to the Nebra findspot suggests a deep connection between the heavens and the humans who lived in eastern Germany. Tracking the Sun and the Moon was crucial for farming, especially aligning Moon-based civil time with the solar, seasonal year. For at least four millennia, European cultures used the Moon and the Sun to track time and probably to conduct rituals, ceremonies, and conquests. The Nebra disk may be the earliest known example of lunar timekeeping in northern Europe; many other lunar symbols came later.

One of the most beautiful, and arguably the strangest, of these is now in Berlin's Neues Museum, where ancient artifacts of central Europe stand next to damaged pillars bombed by the Allies in World War II. It is known as the Berlin Gold Hat.

Dating to the first millennium B.C.E., the giant conical hat seems like something out of Harry Potter, or a Merlin costume for a production of *King Arthur*. And this is probably a legitimate comparison; though its function has been lost to time, the giant hat was most likely worn by a priest or ruler carrying out some kind of lunar ritual. The hat's brim is reinforced with bronze wire, and its dimensions fit an average male head. The wearer was probably in charge of some agricultural or astronomical task, designed to control timekeeping and the structures of civil life. He would have been very powerful indeed. He may have even been viewed as divine.

The hat is mounted on a circular plinth in a darkened room, and its gold is dazzling. It stands thirty inches high, as tall as a desk, and is forged of a single gold sheet. Mounted on its plinth, the hat rises taller than me and looks incredibly imposing. The entire hat is stamped in circular ornaments, which add up to fifty-seven Moons—fifty-seven months, roughly four and a half solar years. The hat was probably used much like the Nebra sky disk. It was probably used to calculate the all-important intercalary month, the leap month, that would bring the lunar year in alignment with the seasons of the Sun. Without the Nebra disk or the ability to write anything down, ancient people in central Europe still found a way to count time, to mark their year, and to use this ability to structure their lives.

Though the Berlin Gold Hat came to the Neues Museum without any provenance, three other similar hats also emerged from the earth, were found by farmers, and now sit in museums. One was discovered in Schifferstadt, in 1835, and appears to have been buried intentionally. The Schifferstadt hat dates to 1400 B.C.E. The Avanton and Ezelsdorf-Buch

Golden Cones, found in France and Germany, respectively, were uncovered without brims. The cones are so similar to the other hats that they probably once had brims, too, and were likely used for similar purposes. All these golden hats date from 1400 to 800 B.C.E., a few centuries after the Nebra sky disk was forged. All of them are hammered from single gold sheets and contain alloys of copper and tin, the most powerful metal of the Bronze Age. Each of them is stamped in golden orbs representing the Moon.

Just like the Mesolithic Scots of Warren Field, these ancient hatters knew they needed to somehow connect the solstices, when the Sun seems to stand still, with the cycles of the Moon. In this way, the Moon would allow them to track the year as well as the seasons.

Warren Field represents the first time people figured this out; the Nebra sky disk may be the first time they figured out how to use the heavens to do it, not just the landscape. The hats are a progression of the same idea.

ALL THESE SKYWATCHERS apparently shared one other thing: They decided to leave no traces of their precious timekeepers.

For reasons we can never know, the users of the Nebra sky disk buried it on the Mittelberg. The realm of the sky disk never grew beyond modern-day Saxony-Anhalt. The area's inhabitants lived in peace for four hundred years; neither Meller nor any other archaeologist has been able to find evidence of major fires, battles, massacres, or any other violence. The villages were not fortified. There were no golden walls like the ones erected in Babylon, twenty-nine hundred miles to the east. The sky disk's owner, or the ruler whom the sky priest served, was part of an agrarian society with a large standing army that protected the land and the people. They controlled the flow of amber, copper, tin, and other valuable materials through the heart of Europe. The Únětice culture had a powerful empire; why didn't they last?

Meller believes they may have been ravaged by plagues from foreign lands, as migration and trade introduced new people, new customs, and new diseases. Their lack of a written language may have even played a role. If the Nebra disk maker learned the secrets of the sky in the lands of Mesopotamia, they certainly did not bring home cuneiform script, the first form of writing on planet Earth. As always, knowledge was power, and for the Únětice, stories were apparently the only way to transmit it. The creator of the sky disk either refused or was unable to share their knowledge. Meller told me he thinks the disk itself, especially the later addition of the solar "boat," is evidence of a slow slide into mysticism. The disk was built as an astronomical device, but in its later years, Meller argues, it was imbued with myth. Ultimately, it was hidden from view, buried in the earth far from the sky it represents, away from the Moon whose phases it was designed to use.

TODAY, THE FINDSPOT of the Nebra disk hosts an observation tower and a nearby museum, hanging above the town like a ship, or maybe a solar barque. It is called Arche Nebra, "Nebra's Ark," and it tells the story of the disk and its people. Bettina Pfaff, the museum's director, took me through carefully constructed exhibitions that depict the various humans to have lived in central Europe, from the invention of farming to the disk and into the Middle Ages.

The rooms are designed to represent the images on the disk—from outside, the walls look like a crescent Moon, but within, their swooping sides display artifacts related to the disk and the people who wielded it. According to Pfaff, the design of the museum building was meant to "set the content free." The museum even tells the story of Westphal, Renner, and the sting in Basel through a puppet show. The curators are careful to not turn the treasure hunters into heroes, Pfaff says.

The disk was buried with several valuable items, including daggers and jewelry. Pfaff points out that it was put into the earth at the end of a prosperous era, but a few centuries before the Bronze Age collapse.[4]

"Maybe the times had been so bad, maybe they buried the most valuable thing they had," she speculates. Maybe its owners laid the item to rest like they would a loved one or a powerful prince, surrounding it with valuable offerings.

Whoever buried the disk made a few alterations before its interment. The left arc seems to have been taken off. This could have been an accident, or maybe not, Pfaff told me. Intentional destruction of vital artifacts occurred often throughout the Bronze Age.[5] The recumbent stone circles of northeastern Scotland were closed off. Warren Field's timber hall was deliberately burned down. Across the Irish Sea, Celtic swords were crushed before being offered to the waters. The list goes on.

"Maybe they meant to destroy it, and bury it forever, so that no one could use it anymore," Pfaff said. The thought was chilling. What were they trying to hide? Or what were they trying to protect?

AFTER LUNCH IN the museum's café, Pfaff took me to the findspot on the Mittelberg hill in the nearby forest. A family of hikers stared at us as we drove up the gravel road. At the top, I climbed an observation tower that Meller's teams of creative architects erected to commemorate the resting place of the disk and its accoutrements. On the ground below, engraved metal railings mark the path of the Sun on the solstices. The tower is situated so that during the June solstice, the setting Sun shines into the tower's central shaft. Sunlight spills through the monument, illuminating its yellow interior.

After climbing fifteen flights of stairs, I emerged onto a platform that overlooked the hills of central Germany. The beech trees were leafing out among the tall spruces, and I could see forever through the Ziegelroda, past Nebra, and to faint hills in the distance.

Below, the hole where the disk was dug up is now covered with a plastic cap. Its endoming bothered some who wanted the hole and its contents re-created so that tourists could see what the diggers saw. But Pfaff and Meller took a more philosophical approach. The covering allows you to see inside, but it is also meant to reflect the sky above, inviting the visitor to imagine the celestial sphere that the disk represented and was designed to echo.

The Moon was not up yet when I climbed down the Mittelberg and into the car with Meller's driver, an easygoing eastern German named Rolf. We took the Autobahn back to Halle, where I boarded a train back to Berlin. In Meller's black Mercedes-Benz, Rolf noted that the Autobahn has no speed limit. "It's okay if I go fast?" he said more than asked. "Sure," I said. He pressed the gas. The car roared to life. The last time I checked the speedometer, the black Mercedes was humming at 224 kilometers per hour, or about 140 miles per hour, which is faster than it is legal to travel on any road where I come from and faster than I could handle. I closed my eyes.

Later that night, I stepped outside my hotel to look for the Moon and the Pleiades. It was early April, and I knew they would be hanging in the sky just as they are depicted on Meller's glorious disk. In northern Germany, the four-day-old crescent Moon was a lazy sickle, a curve toward the bottom right of where the Moon's disk would be. The Pleiades were hard to spot in light-polluted Berlin, but they were there, above the Moon. The following year, 2020, would be a leap year. The disk could have told me that.

○ ˙

THE NEBRA SKY DISK and its attendant museum would both be incredible tourist attractions, lovely archaeological treasures in the German countryside, even if they had no higher meaning. But they are momentous milestones on the journey human consciousness walked with the Moon for more than thirty-five thousand years. In prehistoric human minds, the Moon started out as a fertility symbol, a time counter, and a form of notation. It soon progressed to a new role as a time reckoner, enabling people to orient themselves in time, imagining the future as well as recalling the past. And it was immortalized in the first primitive watch, hammered in gold and used to keep the lunar calendar in sync with the seasons. The Moon's role as a marker of time connects our ancestors to us, through untold millions of lunations. The Moon is responsible for the beginning of time.

BUT AS THE Stone Age collapsed into the Bronze Age, the Moon would become useful for much more than anticipating the seasons. It was the most important feature of the cosmos as humans shook off the heavy mantle of prehistory. The Moon's time-setting abilities meant humans could use it to plan, which meant they could invent. Using its cues, people figured out how to grow enough food that they could stop chasing and foraging. The Moon enabled the beginning of history. The first pieces of human writing owe their existence, and their content, to the Moon's cycles. As the first literate civilizations arose in Mesopotamia and Egypt, the Moon took on importance as more than a marker of time. It became a recorder of events; a predictor of fates; an instrument of might; and a god in its own right. The Moon would ultimately lay the foundations of philosophy and religion, in the East and the West, and set the course of history.

The Moon's dependable timekeeping had helped humans prepare for the future. But its role as a silent, silvery clock in the firmament was soon to be in the Moon's own past.

CHAPTER SEVEN

The Ornament of the Sky

It is the privilege of antiquity to mingle divine things with human.

—Livy, *History of Rome*, book 1, part 7

O Sîn, king of the gods of heaven and the netherworld, without whom
no city or country can be founded or restored.

—The Nabonidus cylinder of Abu Habba (The Dream Cylinder of Sippar),
col. ii.26–32, trans. Paul-Alain Beaulieu

LONG AFTER OUR languages and civilization have been lost to history and archaeologists of the future dig out the British Museum, they might find some of its collections hard to interpret. The looted Parthenon Marbles would just be pretty carved friezes, with no greater meaning behind the *Horse of Selene* or the centaurs. The Rosetta Stone and its inscription in three languages would just be a big granite slab covered with indecipherable scribbles. Countless objects spanning thousands of years of human history would be things of beauty, but of enigmatic purpose.

But there are a few exceptions. Certain artifacts need no semiotic interpretation, because their symbols have meant the same thing throughout time. When something is decorated with a crescent, there is no doubting what that symbol represents. Sometimes more data is necessary to figure out what the Moon signifies or why it's being used, but the crescent itself is unmistakable. What else could it be?

One such artifact displayed at the British Museum might catch the interest of future excavators for a couple of reasons. It contains a remarkably obvious depiction of the Moon, and an equally obvious depiction of worship.

It is a basalt stone, rounded at the top, shaped like a headstone and about as big. It depicts an upturned crescent floating above a man with his right arm raised, saluting. He is engaged in an act most people would recognize as prayer. The man is carved in relief, and he is standing to the left, facing the crescent and other symbols on the right. He wears long, resplendent robes, which are etched in intricate designs indicating that they were embroidered or woven. This would have been an expensive garment, fit for a king. The regal figure is wearing a pointy hat and a neat beard nearly as long as his hat is tall. In his left hand he holds a staff, a symbol that has been used across societies for thousands of years to represent power. The staff has a crescent at its top, echoing the crescent Moon in the sky. Lined up beside the Moon and more distant from the king are a winged disk and a seven-pointed star.

If you visit the British Museum, the label beneath this artifact will tell you that this is the stela of Nabonidus, the last king of Babylon. He is praying to the Moon God Sin, the ever-favorite object of his heart. The stela depicts Sin as the supreme deity, more important than the Sun God—the winged disk—or the morning and evening star, which we know as the planet Venus. This hierarchy reflects a choice that would be Nabonidus's downfall, and that of his entire empire. The king's devotion to the Moon above all else brought about a final end to the finest and most sophisticated culture of ancient Mesopotamia. But the full, glorious story of this man and the Moon would never fit on a museum label.

THE STELA IS just a whisper of the long history, beginning more than five thousand years ago, that humanity has shared with the Moon God, once the most popular deity in the ancient Near East. Though the Moon God has long since faded into history, his influences surround us to this day. The various traditions of his followers—animal sacrifice, devotions, temples where people would gather to worship him, totems featuring his image—would appear in all the organized religion that followed. During part of the Sumerian era, the Moon God was worshipped along with his wife, Ningal, "the Great Lady" and goddess of reeds, and their son, the Sun, making him the chief god of the first known holy trinity.

The Moon God was one of the first gods in human history, if not the very first. For a long time, he was the most important deity of all, and worship of the Moon God laid the groundwork for all the other faith traditions that followed.

AT WARREN FIELD and in the Nebra sky disk, people brought the Moon to Earth for the practical purpose of timekeeping. In the Fertile Crescent,* humans used the Moon to transcend the rhythms of daily life. Through them, the Moon became holy.

The Mesopotamians saw, in the Moon's waxing and waning, the cycles of birth, death, and rebirth. For three nights every month, the sky is Moonless. On the third night, it rises again. The people who built Warren Field and the people who hammered the Nebra sky disk understood that the Moon held cosmological and spiritual power, and they tried to channel some of this power through themselves. But Mesopotamia was the first place where the Moon itself was exalted for its power.

The Moon transformed humans from skywatching bands of wanderers into landowning farmers who created civilization. As far as archaeological records show, this happened for the first time in the southern plains between the Tigris and Euphrates rivers. By 5800 B.C.E., people were living permanently in the verdant Eden that was Mesopotamia (from a Greek translation of "the land between the rivers"). The first cities rose from the plains and were populated by tens of thousands of people. The first was Uruk,† the home of Gilgamesh, which at its height in 2900 B.C.E. probably had eighty thousand residents.

Mesopotamia seven millennia ago was marshy, glacier worn, and supremely fertile, but with unpredictable rain and blazing-hot summers. To solve this problem, people invented irrigation, diverting fresh water into their fields. Once land was worked in this way, it became valuable. People needed a way to protect their land, and to secure some guaran-

---

* The region is not named for the Moon, though it's a nice coincidence. The Fertile Crescent was the cradle of human civilization, a marshy arc that spanned the Nile delta in Egypt, up through the Levant (which includes what is now Israel and Palestine), into southern Turkey, and down into what is now Iraq and western Iran.

† The city's name may have influenced the Arabic word *al-Irāq*, from which the modern country takes its name, though scholarship is split on this etymology.

tee for it if they sold it or gave it to someone else. Law and finance arose; so did writing.

AS PEOPLE STARTED harnessing the forces of nature, they turned to the Moon for aid of a more spiritual sort. Political rulers were able to exploit this early devotion for their own gain and build the world's first empires. As these empires grew more literate and capable through the ages, rites dedicated to Moon and sky worship grew more complex and more intentional. Moon worship required careful observation of the heavens for sacred purposes, more than the civil purpose of timekeeping, and this led to something new: the first stirrings of science. In endowing us with religion, the Moon taught humans both a new means of control and a new form of thinking. The Moon would bring our ancestors closer to a full understanding of the celestial bodies, and of the world around them.

IN HIS YOUTH, Nabonidus would never have expected to become king. "I am Nabonidus, who have not the honor of being a somebody— kingship is not within me," he wrote during his reign, according to a clay cylinder found at the ruins of Sippar.[1] Raised by a mother who worked as priestess to the Moon God, he probably began his career as a bureaucrat before ascending to power in his sixties.

As king, he was evidently uninterested in conquest, instead preferring to study archaeology and architecture. He spent great amounts of time and capital excavating the ruins of his own civilization's forebears, placing special emphasis on the temples they built to the Moon God Sin. He is sometimes called the first archaeologist.

Nabonidus restored temples to the Moon God both in his birthplace, a Turkish city called Harran, and at Ur, one of the world's oldest metropolises. He oversaw the restoration of the wedding-cake-like step pyramid called the Great Ziggurat of Ur, originally built in the late third millennium B.C.E. Nabonidus painted the façade black and rededicated the sacred Giparu, the temple's inner sanctum, for his daughter Ennigaldi-Nanna. He made her High Priestess of the Moon God, just like his mother had been. Ennigaldi built a museum and a school there, in the most luxurious conditions available. Ennigaldi's fastidious treasure collecting, like the elaborately braided hair she wore, connected the king and the princess to earlier dynasties* that had ruled over the great grassy plains between the Tigris and Euphrates rivers.

Thousands of years ago, the ziggurat would have broken the horizon for miles and miles, just like the bonfires inside the stone circles of northeastern Scotland. The ziggurat and its Moon temple would have been a beacon, lighting the night and calling its people before the Euphrates moved ten miles to the east, before the city fell finally into ruin, before the temple and all of Ur were swallowed by the sands of what is now Nasiriyah, Iraq.

MODERN ARCHAEOLOGISTS BEGAN to dig out Ur in 1853, mostly on behalf of the British Museum.[2] Before their excavation, the ziggurat was merely a *tell,* the Arabic word for a mound. Ur's Great Ziggurat was called Tell al-Muqayyar, the "mound of pitch," because its bricks were held together by bitumen, a component of road asphalt. It was probably the first valuable product to spill forth from Iraq's vast oil fields. The oily, durable mortar waterproofed the baked clay bricks, and bound the ziggurat together. Its resilience helps explain why so much of the ziggurat remains, even after four thousand years of reaching for the heavens, even after damage inflicted by invaders both ancient and modern.

The British Museum is full of artifacts like the Nabonidus stela. Much of what comes from the Babylonian civilization, and from the ruins of Ur in particular, were uncovered by an archaeologist named Leonard Woolley. In the Moon God's temple, he found a distant forebear of the museum that now houses his loot. The High Priestess of the Moon God, Nabonidus's daughter Ennigaldi-Nanna, kept what might be the world's first museum of antiquities. The relics tell a profound story of the Moon God and those devoted to him.

Woolley liberated them from the sand in 1927 C.E. He was initially befuddled by the motley collection of objects, which included a bound-

---

* A modern ruler named Saddam Hussein also restored the ziggurat's lower foundations, with the same motivations in mind. During the American-led war in 1991, Saddam parked fighter jets near the ziggurat because he imagined the Americans would not risk destroying the priceless monument. It was still damaged by bomb shrapnel.

ary stone dating to 1400 B.C.E., a statue of a king dating to 2280 B.C.E., foundation bricks from 2220 B.C.E., and an even older piece of weaponry. "The evidence was altogether against their having got there by accident," Woolley wrote of the collection, noting that they instead had "a curious air of purpose."[3]

Some of the relics made sense after Woolley's team of excavators found a barrel-shaped cylinder the size of a soda can. It was etched in four columns of cuneiform writing, humankind's earliest written script, and it contained what may be the world's oldest museum label: "These are copies from bricks found in the ruins of Ur . . . which I saw and wrote out for the marvel of beholders," an ancient scribe had written.

Woolley then found another object, this one a little easier to interpret. It was a carved calcite disk bearing the image of a robed woman with a large nose, long braided hair, and a pointy hat, much like the one Nabonidus wears on his stela. She is followed by attendants who are bringing offerings toward the ziggurat, the Moon God's temple. The woman was also a High Priestess of the Moon God. Her name was Enheduanna, and her father, Sargon the Great, was the king seventeen centuries before Nabonidus. The museum curator, Princess Ennigaldi-Nanna, was a successor.[4]

The two priestesses were separated by more than 1,700 years of history, but they were united in their common heritage and their common cause. Both women were elevated to the EN-ship, or Moon priestesshood, by their fathers, both kings of the lands of Mesopotamia. The EN-ship was one of the most powerful positions in civilization, and the kings exploited that power through their daughters. The Moon God was their shared patron, tool, and idol. Enheduanna and Sargon called this god Nanna, in the language of Sumer. Ennigaldi and Nabonidus called him Sin,* in the language of Sumer's successor, Babylonia. He was the Moon, personified.

---

* Though Ennigaldi and her father would have called the Moon God by the name of Sin, her own name is derived from a more ancient Sumerian spelling of the same deity. Nabonidus was deeply interested in archaeology and his own country's history, and apparently gave his daughter an ancient Sumerian name that means "the desire of the Moon God."

THE IDEA OF Moon and Sun worship probably followed the celestial bodies' original roles in human life. Most early societies have stories about nature gods, because nature controlled their lives; of course you would look to the heavens for aid, from whence comes the life-giving Sun and rain, and the crop-killing hail and pestilence.

For hunters and gatherers, the Moon was literally a guide, allowing people to see at night and to plan their lives. When people got good at growing food and taming animals and built the world's first cities, they continued using the Moon for practical reasons. It shone upon the fields and the city streets. You could watch your grains and your sheep, and you could walk down the road at night to your home or shop. This obvious benefit eventually evolved into a more spiritual connection to the sky and its forces. Maybe the holy Moon was a vestigial reminder of the first farmers' hunter-gatherer heritage; maybe the original stories morphed into myths, and the Moon's light became evidence of beneficence or intentional aid. Early odes to the Moon God are accented with this legacy. They were probably passed down through spoken stories first, and then eventually inscribed on tablets made of clay, the world's first written records. The writing on Sumerian clay tablets was cuneiform, made by pressing a triangle-shaped stylus into wet clay and baking it dry in the sun. Early poems on these tablets revered the Moon God as a kind of holy wrangler, who would watch over the herds of Mesopotamia and ensure plentiful milk, butter, and meat.

The uncredited poem "The Herds of Nanna" tallies up all the cows under his care, some 39,600, apparently, and praises him as "god of living creatures, leader of the land." Nanna was also the blessed purveyor of beer, the most popular drink in Mesopotamia.

"He is able to provide abundantly the great liquor of the mountains, and syrup, and alcoholic drink," the poem reads. "O Father Nanna, be praised!"[5]

Nanna is sometimes presented as a bull himself, in a motif that recurs in all the earliest devotional practices throughout cultures on Earth. The cave painters of Lascaux fixated on aurochs. Ancient Egyptians worshipped Apis, a bull adorned with a full Moon between his horns, as a powerful fertility god. To the east is Shiva, the supreme being in one of the major veins of Hinduism, whose steed is a huge bull. The

Hindu god Krishna, who represents love and compassion, was a cowherd in his youth. In the Old Testament, worship of a golden calf idol is an act of ghastly apostasy.* High Priestess Enheduanna described Nanna this way:

"Shining bull, lift your neck to [the Sun] who ...† in the sky! Your shining horns are aggressive, holy and lustrous. Bearing a beard of shining lapis lazuli, ..., your prince, the mighty sunlight, the lord who ... the true word, who lightens the horizon, who lightens the sky's ... vault."[6]

In poetry and prayer, Nanna/Sin is referred to as "father of the gods," "head of the gods," and "creator of all things." This is how people viewed him long before the first monotheist set out from the Moon City Ur and moved to the Promised Land.

In the time before that story, which comes from the Book of Genesis, early temples were erected to nature gods throughout the Fertile Crescent. One of the very first known settlements was Jericho, today a Palestinian city on the West Bank, in disputed territory near Jerusalem. The earliest relics at Jericho date to an era before humans invented pottery, around 9500 B.C.E., a millennium and a half earlier than the pits of Aberdeenshire's Warren Field. In antiquity, people came to Jericho for two reasons: a freshwater spring and the Moon.

Jericho's spring, now named Ein es-Sultan, was a popular gathering spot for the hunter-gatherer people called the Natufians. Beyond its life-giving waters, a spring is also a potent symbol for people fixated on human fertility, as we know many Neolithic people were. Like other early hunter-gatherer groups, the Natufians are known by the tiny stone tools they left behind. They are called lunates: small crescent-shaped stones used to cut grasses. Natufian hunter-gatherer groups visited the Jericho spring in warm seasons. Around 9600 B.C.E., a period of droughts and cold called the Younger Dryas finally ended, and the Natufians stayed put in Jericho. The oldest city on Earth grew up around these

---

* Exodus 32:1–8, 1 Kings 12:26–30, Deuteronomy 9:16, and other chapters in the Old Testament/Hebrew Bible tell the story of a golden calf idol drawing God's ire.

† The ellipses represent lacunae, the term for missing fragments from the ancient tablets that have been lost to time—or were destroyed by later people.

water seekers. Befitting the spring's connection to fertility rituals, Jericho became a pilgrimage site for worshippers of the Moon.

Scholars have a few theories for the origin of the city's name—some say it derives from a word meaning "fragrant," describing its abundant flowers—but the Palestinian government's tourism office describes Jericho as "the City of the Moon." Jericho was an early center of worship for a Canaanite god named Yarikh, a god represented by the Moon. People traveled to the city to visit a temple to his honor. This may also explain the origins of other proto-cities of the third millennium B.C.E. The temple probably came first, and a city stirred to life in the buildings erected around it. "Ur of the Chaldees" in Sumerian cuneiform script is 𒀕𒆠, rendered as URIM[ki] or UNUG[ki], which translates to "the abode of Nanna."[7]

MESOPOTAMIA BY THE third millennium B.C.E. was a hodgepodge of city-states, including Uruk, Ur, Kish, and Akkad, the ruins of which were lost to history. Inhabitants of these city-states fought one another for centuries over the best lands and water access between the rivers. King Sargon hailed from Akkad and became the first to unify the city-states and the first to create something resembling an empire, controlling people through shared bureaucracy and religion. He conquered Ur in 2334 B.C.E. with an army that was stronger than previous Sumerian kings', in part because his troops had bows and arrows, a rare sight in a land without many trees.[8] Sargon's domination was a real feat given his humble beginnings. His autobiographical origin story tells of a baby born to an unknown father and a priestess mother, who gave birth to him in secret, put him in a reed basket sealed with tar, and cast him into the Euphrates.*

When the fighting was over, Sargon needed a way to unify his realm through means that were stronger than arms. Bureaucratic control can have more staying power than military might—the reed stylus is more powerful than the sword, in other words—and Sargon achieved this in a few ways. He standardized weights and measures, which were

---

* Doesn't this story sound familiar? Sargon's birth tale predates that of Moses by eight hundred years.

previously different from place to place. He levied taxes, and he kept members of old Sumerian ruling families as members of his court. He initiated a bureaucracy beyond anything developed beforehand in Sumer. But his most important bureaucrat was his daughter, the well-coiffed priestess Enheduanna. "Daughter" may have been an honorific rather than signifying Enheduanna was Sargon's biological child, but either way, Sargon must have trusted her greatly. On his behalf, she used the Moon to unite the gods of Akkad and the gods of Sumer, solidifying the first empire on planet Earth.

ENHEDUANNA'S VERY NAME was the first step in this process. Her original name was probably Semitic in origin, but changed to a Sumerian form to placate the people she now aimed to unite. Enheduanna is a nom de plume: ⽥ (EN) translates to "lady, High Priestess" combined with ⽥ ⽥ (HEDU), which means "ornament." Followed by ✳ (AN), the sky god, and ⽥ (NA), her name means "High Priestess of the ornament of the sky," of the Moon.

AKKAD INCLUDED THE Moon God in its greater pantheon, but Akkadians were more devoted to the goddess of love. The Sumerians called her Inanna.* The Babylonians would later call her Ishtar. In Greece, she was Aphrodite, and in Rome she was called Venus, for whom the second planet from the Sun is named.

Yet she was no Moon. Nanna was the most popular god of Sumer, and in Sumer he must be exalted. A king who wanted to run Ur needed to venerate Nanna, so Enheduanna was installed as the Moon God's High Priestess and technically as his wife on Earth. She knew she had to honor Nanna, but she also knew the great power she could derive from introducing Ur to Inanna, too, and building a connection between the two deities, and therefore the two city-states. She accomplished this in a poem called "The Exaltation of Inanna."[9]

---

* The name looks similar to Nanna because our transliteration of cuneiform is limited. The first writers on planet Earth didn't have a sophisticated alphabet; just you try coming up with an original set of characters that can convey the entirety of human experience.

At some point during her forty-year tenure, Sargon's grandson Naram-Sin became king of Sumer. Some Ur residents rebelled against the Akkadian leaders and challenged Enheduanna's priesthood. Kicked out of the sacred Giparu, she went into exile and wrote at length of the miseries she suffered, cut off from her sanctuary and unable to perform her holy duties. Without a direct line to Nanna, she appealed instead to Inanna for help. Finally, according to her own writings, the gods answered her plea and she returned to the Giparu. Her tale of woe, written in verse, was an extremely astute narrative choice. She successfully manipulated her story of suffering to turn the focus toward Inanna, not just Nanna. Enheduanna probably proposed declaring that Inanna was Nanna's daughter, as she is later described in Sumerian poetry. This relationship would have justified the goddess's power over Ur and southern Sumer, while maintaining Nanna's supremacy.

The gods are further glorified in Enheduanna's so-called Sumerian Temple Hymns, forty-two verses about the holy places throughout the lands of Sumer. They are not the world's oldest religious text—that is the *Epic of Gilgamesh,* the story of the king of Uruk and his wild friend, which contains the earliest references to the Flood—but Enheduanna's hymns began to change the way people in the Bronze Age viewed their gods. These hymns made the gods more interesting and more human. They had thoughts and cares and went on adventures. Inanna became Queen of Heaven. In Enheduanna's telling, Inanna visits the wise old patricidal god Enki* while he is drunk and tricks him into giving her the gods' collective cosmic powers, known in Sumerian as the *me.* This shows Inanna's cunning as well as her sheer power. Nanna in his many manifestations became a deeper, more well-rounded character. The gods seemed more invested in the lives of everyone, not just certain Sumerians or Akkadians. The Moon priestess's poems are sensual, erotic, playful, humane, and deeply passionate. Through her writing, humanity tried to make connections between heaven and Earth for the first time. Bronze Age humans narrowed their focus of worship, moving from the abstract affairs of nature gods toward their direct

---

* The same god who killed his father, Apsu, maker of heaven and Earth with Tiamat.

involvement in the lives of people. Thanks to Sargon's daughter, the most interesting gods now had more in common with their worshippers.[10]

The temple hymns were probably written to unify the lands Sargon conquered. Though a few versions probably existed earlier, Enheduanna claimed credit for penning them, and in doing this she is recognized as the first named author, the first person to ever byline a piece of writing. She herself wrote: "The compiler of the tablets was Enheduanna. My King, something has been created here that no one has ever created before."[11]

The forty-two temple hymns were copied and recopied for more than two millennia, longer than the recorded history of the New Testament. Enheduanna's evocative, ardent style laid the foundations for the liturgical works of Judaism and early Christianity. In "The Great-Hearted Mistress" (also sometimes translated simply as "A Hymn to Inanna"), the poet writes:

> *You are magnificent, your name is praised, you alone are*
> *   magnificent! ...*
> *My lady ... I am yours! This will always be so! May your heart be*
> *   soothed towards me! ...*
> *Your divinity is resplendent in the Land! My body has experienced*
> *   your great punishment.*
> *Bitter lament keeps me awake with ... anxiety. Mercy, compassion,*
> *   care,*
> *Lenience, and homage are yours, and to cause flooding, to open*
> *   hard ground and to turn*
> *Darkness into light.* (lines 218, 244–53)[12]

Abject worship like this has come down through the ages. You can hear these plaintive lines echo in any church today: *The Kingdom and the Power and the Glory are yours, now and forever.*

UNITING THE GOD OF UR, the Moon City, with the most important Akkadian goddess gave the Akkadian dynasty legitimacy and secured the priestess's own position. By connecting Inanna and Nanna, Enheduanna maintained her grasp on power, and so did her family. She was so successful at uniting the South and North that later kings would appoint their daughters to the position of High Priestess of Ur long after Sargon's dynasty faded into history.

A woman with so much influence would have loomed pretty large through the centuries. So perhaps it's no wonder that Ennigaldi, daughter of King Nabonidus, kept relics from Enheduanna in her personal museum. The earlier priestesses of Ur may have been honored as heroes and icons, in the way modern Catholics pray to saints for intercession. Woolley, after pulling artifacts out of the ground, remarked that earlier priestesses were probably honored for political reasons, just as the gods themselves were. "Whether the god was Jehovah or Nanna mattered little" to later rulers, Woolley wrote. "The king's purpose was to placate his people by subsidizing their particular forms of worship."[13]

ENHEDUANNA'S CLEVER HYMNS may have served as the mortar linking the disparate bricks of Sumerian society. But it would not last. "Old things are passed away; behold, all things are become new," as another holy author would later write in a letter.[14] In the centuries after Enheduanna's tenure, Ur was besieged by rebellion and invasion and a series of impotent, short-lived rulers. "Who was king? Who was not king?" a document called the Sumerian King List asked ruefully. By the turn of the second millennium B.C.E., the third dynasty of Ur was suffering from famine. Centuries of irrigation had started to take a toll. The waters of the Tigris and the Euphrates are slightly salty, and over time the canals that fed the fields also deposited salt, changing the mineral makeup of the soil. Crops began to fail.

In 2004 B.C.E., four millennia before U.S. Marines toppled a forty-foot statue of Saddam Hussein in Baghdad, just fifty-five miles north, a Babylonian king named Ibbi-Sin was faced with a coup. Marauders broke down the walls of Ur, burned the palace, leveled the Moon God's temple, and brought the Sumerian era to a final, staggering end. Poems described the city's woe and, more ominously, the failure of Nanna and his wife to protect it.

Father Nanna . . .

Your song has been turned into weeping before you—how long will this last?

Your city weeps before you as its mother. Ur, like a child lost in a street, seeks a place before you. . . .

Your brick-built righteous house, like a human being, cries, "Where are you?" . . .

How long will you stand aside from your city like an enemy?[15]

The most important god in the Sumerian pantheon, the oldest god of gods, had lost some of his potency. The nature gods of the Near East were shaken for the first time. They would finally succumb fifteen centuries later, when the Moon God failed to save his most ardent patron, the last Semitic king of Babylon: Nabonidus.

◯ ˙

HISTORY IS WRITTEN by the victors, so the Moon God has long since faded from glory in favor of the God of Abraham, Isaac, and Jacob. The Torah, the Christian Bible, and the Quran mention the Moon God mainly to condemn his worship. The holy books do not mention Ennigaldi or Enheduanna or the other uncountable women who made the Moon God paramount in the thousands of years before the God of Genesis revealed himself. But a straight line connects the Moon to the God of Genesis, the God of Abraham, the God most people on Earth worship today. The line runs through the first monotheist, Abraham himself, and through his hometown: Ur, the original Moon City.

In the books of the Bible, Abraham is the progenitor of Judaism, Christianity, and Islam. In Genesis, God orders him to leave Ur and make his way to Canaan, where he is to become father of many nations. He is the patriarch of the Twelve Tribes of Israel, through his grandson Jacob. In Islam, he is known as Ibrahim, the "friend of God" and the father of Ishmael, the ancestor of the Arab nations. No archaeological record of him has ever been found, but according to multiple faith traditions, he grew up in Ur. If that is true, Abraham—born as Abram—would have grown up surrounded by people who at least paid lip service to the

Moon God, including his father, Terah, and his wife/half sister, Sarai. Her name is the Akkadian version of Nanna/Sin's wife, the goddess Ningal. Given this legacy, it makes sense that, according to Genesis, God renames them Abraham and Sarah[16] when he makes his new covenant. Their new monikers symbolically ditch Ur and the Moon, and add the new syllable "AH," the first syllable of the name of the God of the Israelites in Genesis (YHWH).* The parable is a way to erase an older, well-loved deity and replace him with a new, singular one, who would eventually overcome all others. According to Genesis, Abraham and his family leave Ur, then take a detour to Harran, before heading to the Promised Land. Alongside Ur, Harran was the most important center for worship of the Moon God from the third millennium B.C.E. until the rise of the Persian Empire. It was where Nabonidus was born and where his mother worked as High Priestess to the Moon God.

King Nabonidus's story also comes down to us from the Hebrew Bible. We meet Nabonidus in the Book of Daniel, where his name is confused with that of his predecessor Nebuchadnezzar II—the Bible's bad Babylonian king, who sacked Jerusalem in 587 B.C.E. and burned the Temple of Solomon. This conflation happened because whichever scribes wrote down the story of Daniel consistently avoided mentioning Nabonidus by name. The error carried through the centuries in part because so many classical scholars celebrated Nebuchadnezzar, attributing Babylon's greatness to that king's long reign.[17] But historical accounts of the time, as well as a long record of cuneiform texts known as royal inscriptions, show that Nebuchadnezzar's successor Nabonidus is the actual "mad king" who goes wandering in the wilderness living among oxen, as the Bible tells it. Historical records and several clay cylinders, now in the collections of the British Museum and Berlin's luminous Pergamon Museum, round out Nabonidus's biography.

---

* Modern usage spells this name as Yahweh or Jehovah, but these are really not names, just ways to spell the words of the name God gives himself: YHWH, according to the story in Genesis 15:7. This four-letter moniker was also later known as the *Tetragrammaton* in Greek.

NABONIDUS BEGAN HIS career as a civil servant, probably a prefect in Babylon, something like an unelected city alderman. At some point during the reign of Nebuchadnezzar II, Nabonidus was working as a military officer. He helped broker an important cease-fire during a conflict between the Medes and the Lydians in Asia Minor, according to Herodotus. Some tablets dating to the reign of Nebuchadnezzar suggest Nabonidus could be abrasive and cantankerous; in one account, he ordered the beating of a man who had inquired about a robe adorning the statue of a god. But he was by all accounts pious, interested in archaeology, and devoted to the Moon. By the time Nebuchadnezzar died, in 562, Nabonidus was probably a trusted member of the court. Chaos followed the death of Nebuchadnezzar, whose son and grandson both ruled briefly before being assassinated. Ultimately, perhaps via a coup orchestrated by his adult son, Belshazzar, the pious bureaucrat Nabonidus ascended to power.

The new king dearly loved the Moon God, his mother's favorite deity, to whom she dedicated herself as priestess. The Moon God held great influence over the son long into his adulthood. In the second year of his reign, he rededicated the Temple of Sin in Ur during a lunar eclipse, and installed Ennigaldi as High Priestess there. Cuneiform cylinders from the reign of Nabonidus tell of his piety and his many architectural works in the Moon's honor, including the restoration of the temple where his mother worked. Other scattered relics throughout the Middle East fill in the story of Nabonidus and his Moon. In 1881 C.E., an Iraqi Assyriologist named Hormuzd Rassam uncovered the ruins of the Temple of Shamash, the Sun God, in the ancient remains of a city once called Sippar. He found temple treasures and a small clay cylinder covered in tiny cuneiform script. The cylinder, dating to the sixth century B.C.E., is now in the Pergamon Museum. The Canadian Assyriologist Paul-Alain Beaulieu translated the cuneiform in the 1980s C.E., and found that it contains this prayer, written on behalf of King Nabonidus:

May the gods who dwell in heaven and the netherworld constantly praise the temple of Sin, the father, their creator. As for me, Naboni-

dus king of Babylon, who completed that temple, may Sin, the king of
the gods of heaven and the netherworld, joyfully cast his favorable
look upon me and every month, in rising and setting, make my omi-
nous signs favorable. May he lengthen my days, extend my years,
make my reign firm, conquer my enemies, annihilate those hostile to
me, destroy my foes.[18]

But by the time Nabonidus ascended to the throne in his sixties, with
a son, Belshazzar, in his forties, the Moon God was less popular than
Babylon's Sun-centered agriculture god. Marduk's priests made Nabonidus miserable, and the king complained about not being able to venerate Sin, his mother's god. He ultimately grew so unhappy that he made
his son co-regent, left him to rule in his place, and departed Babylon
altogether. With his daughter Ennigaldi safely ensconced as the Moon
God's priestess, and his son, Belshazzar, in charge in Babylon, Nabonidus went on a walkabout in Arabia.

Scholars think he may have had a few motivations for this absence,
including an attempt to control trade routes through western Arabia.
There is also evidence he suffered from some kind of illness, maybe a
skin malady. The Dead Sea Scrolls, discovered at the Qumran Caves in
the West Bank between 1946 and 1956, preserve a work called "The
Prayer of Nabonidus," in which the king is said to suffer from an "evil
ulcer." In Babylonian medicine, several skin conditions, from psoriasis
to leprosy, were interpreted as divine punishments from Sin.[19] Maybe
Nabonidus fled his city to avoid infecting others; maybe he fled to the
dry desert for relief; perhaps he rebuilt temples to his favorite god in
desperation for a cure. These stories inspired the tale of the "mad king"
in the Book of Daniel, in which Belshazzar's father (really Nabonidus)
lives among animals for seven years before reaching an understanding
with the God of Israel.

Nabonidus himself wrote of his disgust with the Marduk cult, and his
frustration at being unable to worship as he pleased after his subjects
disregarded Sin's rites.

"As for me, I removed myself out of my city Babylon and [I proceeded]
on the road to Tayma. . . . During ten years, I went back and forth among
these cities and did not enter my city Babylon," Nabonidus complained.[20]

He brought his army to places like Tayma, Dadanu, Yadihu, and other Arabian cities, and visited his birthplace, Harran. The armies of Babylon and the Medes had marched on Harran in 610 B.C.E. and despoiled its venerable Sin shrine. The residents of the city viewed the desecration as a result of divine abandonment—just like Nanna had abandoned Ur in 2004 B.C.E.—and believed that restoring the temple would be a way of restoring Sin. So Nabonidus went home and rebuilt the ancient Temple of Sin, Ehulhul, or "house of joys."

In May 1906, an archaeologist named Henri Pognon found the remains of the temple in the floor of a twelfth century C.E. mosque. Stelae from the temple had been used in the mosque's floors so that Muslim worshippers would walk upon the idol of the infidels on their way to prayer.

Pognon was trained as an epigraphist, specializing in ancient writing, and, while exploring a mound near Harran, he came across a damaged stela with the remains of four columns of cuneiform inscription. His published translation showed that the author of the text was a devoted worshipper of the Moon God Sin. The second column described, in great detail, the reconstruction of the once-famous Sin sanctuary at Harran. The inscription tells how King Nabonidus restored the temple, and references him as "the son, the offspring of my heart."[21]

At first, Pognon reasoned that whoever wrote this must have been a priest of the temple, who held great affection for his benefactor; calling the king of Babylon his son might have been a term of endearment. He later found three more stelae that confirmed the much more interesting truth: The stelae represent a biography of Nabonidus's mother. They tell the story of how Sin appeared to Nabonidus's mother, Adad-guppi, in a dream, promising her that her only son would be called to kingship and would rebuild the temple at Harran. Another inscription describes her death, apparently at the age of 104, and the ceremony Nabonidus held in her honor:

I am the lady Adda-guppi', mother of Nabium-na'id, king of Babylon, votaress of the gods Sin, Ningal, Nusku, and Sadarnunna, my deities. . . . My blessings, the goodly things which they gave me, I too by day, night, month, and year, gave back to them. I laid hold on the hem

of the robe of Sin, king of the gods, night and daytime my two eyes
were with him, in prayer and humility of face was I bowed before
them (and) thus (I prayed), "May thy return to thy city be (vouch-
safed) to me, that the people, the black-headed, may worship thy
great godhead."[22]

May the people—the descendants of the Sumerians and the inheritors of Ur, "the black-headed"—return to the City of the Moon God to
worship him.

Nabonidus made good on this promise, but it would come at a cost:
his kingdom for the Moon.

BY THE EARLY 540s C.E., King Nabonidus was at best unpopular. While
away in Arabia, he made a fateful decision not to come home to Babylon
for the vital New Year's festival held in spring. During these important
annual rites, the king would accompany an effigy of Marduk through
Babylon's splendid Ishtar Gate. The city walls were decorated in lapis-
blue tiles, embedded with daisy rosette mosaics, and guarded by strid-
ing dragons and ferocious lions. The gorgeous blue tiles of the
Processional Way, leading to the glorious gate, were the backdrop for the
New Year's festival. The ceremony affirmed the king's right to power by
simulating the transfer of kingship from heaven to Earth, with Marduk
as the shepherd. You can see parts of the walls today, lining a corridor in
the Pergamon Museum, not far from the prayer cylinder of Sippar.

Nabonidus could never bring himself to honor Marduk, even to accept the transfer of kingship from heaven, and sacrificed his secular
duties in favor of his sacred honor. Though later accounts likely exagger-
ated how much this bothered people, many Babylonians probably were
offended and even outraged. In the east, a ruler named Cyrus saw an
opening.

By 540, the Persian king Cyrus was already sending forces along Babylon's borders, sniffing out territory he might want to invade. Nabonidus's army held them off, aided by a forty-foot moat and double walls.
Babylon's residents had enough food and water to outlast any siege,
even up to twenty years, according to the Greek historian Xenophon,

and maybe this made them complacent. Cyrus and his men decided to seize the opportunity presented by a weakened, unloved, and mostly absentee king. In October 539 B.C.E., Cyrus made his move. His men dug trenches along the Euphrates and diverted the river's waters so they could march under the walls of the city. They crept in through the mud and sang like drunken revelers, escaping notice until they reached the palace.

Inside the palace, the acting king, Belshazzar, was hosting a great feast. According to the legend in the Book of Daniel, he imbibed using precious vessels that had been stolen in the destruction of the Temple in Jerusalem* a few decades earlier, illustrating his unholy debauchery. In the Bible, the nature of the feast is vague. But the date was the sixteenth day of the month Tišritum, a feast day and very possibly a celebration in honor of the Moon God.[23] The Moon was huge and bright, a harvest Moon that rose late and orange through the darkness.[24]

As the revelers partied, a disembodied hand appeared and began to write on the wall in a mysterious language. Belshazzar, terrified, called for someone who could read it. Daniel, an exiled Jewish youth, came over. He read the Hebrew: *"Mene, mene, tekel, upharsin."*

"God has numbered your kingdom and put an end to it. You have been weighed on the scales and been found wanting. Your kingdom has been divided and given to the Persians."

That night, the forces of Cyrus stormed the palace and killed Belshazzar. Historian William Shea suggested in the 1980s C.E. that Belshazzar likely heard the Persians had defeated his father at Sippar and hosted the banquet for his own coronation.[25] But nothing is known for certain of the fate of Nabonidus. He may have gone into exile; he may even have been fortunate enough to live out his days in quiet piety, praying to the Moon every night.

---

* In the Bible, the Babylonians under Nebuchadnezzar sacked Jerusalem and destroyed the Temple of Solomon in 587–586 B.C.E. The stolen vessels may have included the famous Ark of the Covenant, which 1 Kings 8:1–10 says was stored in the Temple's inner sanctuary. The gilded box, said to contain the Ten Commandments tablets, disappeared during the destruction of the Temple. No record of its whereabouts exists, unless, perhaps, it is in a wooden crate in U.S. government storage.

WITH APOLOGIES TO Belshazzar, the Persian takeover was a mostly bloodless coup; fed up with their absent ruler, the residents of Babylon apparently didn't put up much of a fight. What's more, Cyrus took pains to placate the people and ensure he would be greeted as a liberator. Babylon was an immense city, and clay tablets were not exactly newspapers or Twitter feeds, so it took a while for word to spread.[26] "It is said when it was captured," Aristotle wrote later, "a considerable part of the city was not aware of it until three days later."[27]

WITH CYRUS'S UNQUALIFIED victory over Babylon, the power center of civilization moved eastward. The plains between the two rivers were now ruled by the Persians, a neighboring empire from the mountains. The final fall of the Moon God would not take place for another thousand years, until a messenger of God named Muhammad would destroy his likeness inside another temple, called the Kaaba. But the golden age of Mesopotamia was over. And the defeat of Belshazzar and Nabonidus signified a major cultural shift: the first major defeat of the nature gods.

The fall of Babylon dealt an especially devastating blow to the most powerful nature god, Sin. If the Moon God could not protect his most powerful disciple, a king personally devoted to his name and his sacred honor, how could Sin possibly help the meek and downtrodden? What good is a shepherd who cannot protect his flock? In losing Babylon, Nabonidus lost much more than his own kingship. He forfeited the divinity of nature. The Moon, the Sun, the sky, the waters were all exposed as weaklings, controlling just a few aspects of the natural world.

When Cyrus took Babylon, he made another crucial decision: He freed the people of Judea, allowing them to return to Jerusalem and rebuild their temple. In sparing the Jews, he spared their story and their God. The God of Abraham rose in popularity after the fall of Sin. This God sat above all natural forces and all heavenly bodies, orchestrating them like a great conductor. He *built* the natural world in just a week. And he acted alone. No separate, lesser gods helped him or even fought him. No one tried to rule alongside him. No one tricked him and tried to steal the *me* from him. It was just him, Elohim. *I am that I am,* he said.

The word of this God spread across the varying cults of the Near East throughout the next five centuries, and it continued to spread after a philosopher from Bethlehem began teaching that he was this God's son. The glory days of the Moon God were at an end.

MEANWHILE, OTHER CIVILIZATIONS from Egypt to Carthage caught up to the heretofore unparalleled brilliance of the area we now call Iraq. Rome was growing in size and power. And the beginning of democracy was taking hold in Athens. Instead of worshipping nature, trying to placate it and please it and beg it for its intercession, humans and their gods began to transcend nature. So did human ideas. As the nature gods became more obscure, the Moon became a tool for thinking rather than just for subjugation. Though Babylon itself fell, its astronomical records, kept as a form of prayer, would become the first example of scientific evidence observed for its own sake. The Moon was becoming the central feature in a new project of intellectual curiosity.

$$\bigcirc \,{}^{\bullet}$$

DURING HIS RULE, Nabonidus devoted a great deal of energy and re-sources to understanding the messages of the heavens. Cylinders found in the ruins of Sippar record royal orders from Nabonidus to pay beer and food rations to scholars studying the skies. The scholars had their own place to work, called the *bit mummu,* which the Nabonidus expert Paul-Alain Beaulieu terms a "temple academy." Though Nabonidus revered the Moon, there was a more practical reason for his careful atten-tion. And it led to something Nabonidus could never have expected.

Scholars under his orders would study the Moon to better predict eclipses, which were viewed with great superstition. How could a blood-red Moon, or worse, the disappearance of the Sun, be perceived as any-thing but an ill omen? During an eclipse of the Moon, Sin is stricken with grief, "covered in mourning." The word for "lunar eclipse" in the Akkadian language can be written with the logogram 'IR—the same word that means "to cry." Though most young societies have viewed eclipses unfavorably, perhaps none prepared for them so fervently as the people in Babylon, the most sophisticated early society.

Forecasting the next eclipse with mathematical precision allowed kings to start preparing the necessary protective rituals, ensuring they had enough beasts to slaughter, enough incense to burn, enough trustworthy people in court to carry everything out. Precautions included beating a copper kettledrum at the temple gates and shouting "Eclipse!" while people sang lamentations, and an elaborate ritual called *šar pūhi,* the "substitute king ritual."[28] If the sky priests predicted the worst type of lunar eclipse—for instance, if Jupiter would be invisible while the Moon was drenched in blood—the king would disguise himself as a farmer and hide. Some other person would be chosen as a regal doppelgänger and would be dressed like the king while a priest recited the dark omen foretold by the eclipse. The ritual would end with the death of the substitute—fulfilling the prophecy of regicide, while sparing the actual rex.

Nabonidus, his daughter Ennigaldi-Nanna, and their astrologers carried out these rituals for religious reasons. They wouldn't have thought of themselves as scientists, surely not in our modern sense of the word. But the scientific enterprise was nonetheless the result of their devotion. Their work ultimately became the legacy of Babylon. From the dutiful records of celestial omens, the reality of the heavens emerged. This reverent record-keeping was about charting the movements of the actual Moon across the sky, and the result was the beginnings of mathematical astronomy, the first exact science. This work at the Moon City of Ur and in Babylon reached its apex during the reign of Nabonidus.

"'Natural phenomena' became objects of study not in spite of their being products of divine agency and will, but precisely because they were physical signs of divine agency and will," writes the Assyriologist Francesca Rochberg.[29] Though their original motivation may have been holy, the knowledge these skywatchers obtained was secular, and it was for everyone. The sky priests' records started out as simple lists of stars; progressed to charting and acknowledging the relationships between celestial events, and correlations between heavenly and Earthly events; and finally became a way of drawing inferences about these events. The texts that compose the Babylonian compendium called MUL.APIN, the one whose charts and Moon positions show the validity of the Nebra sky disk, is one of the earliest records of science in human history.

THE UNFOLDING OF this new form of science—call it knowledge from heaven—took place in many locations at the same time. In China, astrologers looked to the heavens for divine inspiration and authority, to ensure imperial security, and to better predict future events. And just as in Mesopotamia, Chinese astrologers began to understand real nature as a result.

Early astrology led to a watershed moment in the history of human thought. The Moon and the planets were being referenced and observed not simply because they were divine, not just because they were representative of change, but because their movements were *information*. And information can be used.

The entire body of cuneiform mathematical work, all the records on the clay tablets from three thousand years of Mesopotamian civilizations, are quantitative: numerical records and accounts. They are predictive, because their whole purpose was for sky priests to order the heavens and plan accordingly. And they are empirical, in the modern sense of that word, in that they are based on firsthand observation. The disciples of the Moon God, toiling in his honor, gathered more fundamental truths than they could have expected, and people began to understand the heavens like never before.

"The phases of the Moon, the expression of its image in the sky, is really a good tool for thinking," Willis Monroe, an Assyriologist at the University of British Columbia, told me. "For Mesopotamia as a whole, I think you could really say this is very much what the intellectual project is about. It's about watching the world around them and finding these patterns. For astrology, the patterns are real. They're not just things happening in the world. They are very real patterns that can be studied, and the Moon is far and away the most obvious of these patterns."

BY 500 B.C.E., the residents of planet Earth were busy building something resembling the world as we know it. Huge numbers of people lived in big cities full of waterworks and fortresses and temples. Farmers did an excellent job providing food for hungry and growing populations. Some people quit farming and went into other trades. People invented money. Some journeyed hundreds of miles along the ancient Silk

Road, to trade for metal, clothing, food, animals, trinkets. The Olympics started. The study of history began. And the very early stages of philosophy were taking root from Greece to China.

Nearly 2,500 years had gone by since the people of Mesopotamia invented religion. Now the Persians, conquerors of Babylon, were the leading experts on the daily and nightly happenings above. They made the most advanced measurements of the heavens any humans had achieved yet. Their astrological omen tablets show a detailed understanding of the Moon's motions, eclipses, the movements of planets and stars through the constellations, and how the Moon interacted with them. These astronomers—or call them astrologers, their goal was the same—gave us the signs of the zodiac that we still use today, by dividing the sky into twelve zones and assigning each section the name of the most prominent constellation within it. Do you know your zodiac sign? Studies show that 90 percent of Americans do, when given the list of twelve options.[30] We can thank the Babylonians for these.

The sky priests of Mesopotamia also figured out the eclipse cycle. Earth's solar and lunar eclipses repeat the same geometry every 6,585 days—18 years, 11 or 12 days depending on leap years, and about 8 hours. Think about the precision and perseverance it would take to have figured this out three millennia ago. Before paper, before pen, before what we would consider arithmetic, Babylonian astronomers counted out what we now call the Saros cycle. They did all this work under the aegis of the king, but the records left behind could be interpreted by anyone. And early Greek scientists relied on them to make astonishing new observations about the cosmos and how it functions.

Across the Arabian Desert, in a land to the west of Mesopotamia, an immigrant from the Persian Empire would become one of the first people to use this information for something entirely new: the earliest stirrings of modern science.

CHAPTER EIGHT

The Voyage of Discovery Begins with the Moon

The Sun imparts to the Moon her brilliance.

—Anaxagoras of Clazomenae as recorded by Plutarch, *De Facie*[1]

ON THE NIGHT of August 20, 2017, I sat studying Google Maps for the best way to follow the Sun. The St. Louis forecast for the next day was dicey, and I could not bear the idea of missing the "Great American Eclipse," as news media, myself included, were calling it.

The Saros cycle determined by the Babylonians predicted that the Moon would completely block the Sun on August 21, 2017, resulting in two minutes of daytime darkness across a swath of the continental United States—including my own backyard. But I did not trust St. Louis weather. I canceled plans with friends, woke with my family well before dawn, and drove toward clear skies over Paducah, Kentucky.

The previous eclipse visible across the contiguous United States was in 1918, so Americans had reason to be excited.* It is a rare thing to witness the Sun disappear from the sky, with only its atmosphere, called the corona, visible surrounding what appears to be a black hole. Humans throughout time have marveled at this sight in abject terror, awe, ecstasy, euphoria, transcendence, hope, despair, and infinite variations of those feelings—and this remained true long after people figured out

---

* Another total eclipse crosses the United States on April 8, 2024, and after that, August 2044.

what eclipses were. I knew what to expect, geometrically speaking, but I was excited to find out how the total solar eclipse would make me feel.

WE FOUND A spot on Paducah's riverbank to wait for the Moon to slide into position. I noticed as the sky gradually dimmed to a steely yellow cast. Sunlight through the leaves along the Ohio River fell on the ground in crescent form, like little Moons scattered on the ground. Color seemed to drain from the world. The rooftops were strangely yellowed, as though veiled by smoke, only there was no smoke, and shadows were sharply defined. I kept checking my watch. Then, finally, at 1:22 P.M. central time, the Sun disappeared. In the instant before "totality," as the full eclipse is known, a dazzling beam shone through one of the Moon's craters, causing what is known in eclipse circles as the "diamond ring" effect. The sky turned to twilight, and I stared at the Moon.

The crowd on the banks of the Ohio cheered; some wept. My older daughter, then two, protested that she didn't want to go to sleep. I laughed as the katydids began their dusk-hour clacking. In the Sun's place was a halo of fire surrounding a black spot in the sky. The Sun's disk was gone behind the Moon, because in an incredible celestial coincidence, while the Moon is about four hundred times smaller than the Sun, the Sun is about four hundred times farther away from us.

I expected to feel some combination of fear or exultation, maybe a sense of cosmic kinship, as I stared at the corona. Instead, I felt a giddy sense of friendliness.

The Sun's outer atmosphere was pale, but not ghostly like I had imagined, not spectral like the full Moon on a cold December night. It was quiet, empyrean light. I imagined it reaching gently toward me, which is, in fact, what it was doing. The corona gives rise to the solar wind, and we are all swaddled in it.

With a few seconds left before the diamond ring reappeared, I wrenched my eyes away, back to my family. The sky brightened. Totality was over.

Around noon 2,495 years earlier, the shady groves around Athens, Greece, began casting weird shadows, too. Crescents dappled the streets and the sky dimmed. This eclipse, on February 17, 478 B.C.E., was different; the Moon was more distant from Earth, at apogee, so it was unable to block the entire Sun. A small ring of orange was visible around the Moon's black disk, causing what's called an annular eclipse—still beautiful, still terrifying to behold.

The citizens of devastated Athens, which had been pummeled by the Persian army two years earlier, might have cried out, just like the people around me in Paducah. But this ancient Athenian eclipse was different from any before it. That day, perhaps for the first time in the history of human thought, someone was looking up who understood what was happening.

A twenty-two-year-old Persian refugee named Anaxagoras wrote down what he saw, determined to find out more about where the eclipse shadow fell. Like his scientific descendant Alfred Wegener 2,400 years later, Anaxagoras started by asking his neighbors. In the weeks and months after the eclipse, he went down to a port city called Piraeus and interviewed traveling merchants and sailors, asking them what they witnessed and compiling a narrative. He learned that the eclipse shadow, called the umbra, covered the entire Peloponnese. It covered Athens and spread all the way to Anaxagoras's home city of Clazomenae, in Asia Minor. But beyond the realm of the Greek world, it was not visible. This was a shocking revelation at odds with all other knowledge about eclipses up to that point. The darkness of the eclipse was limited.

Anaxagoras had already developed several novel ideas concerning the size of the Moon, its distance from Earth, its nature, and its place in the heavens. His eclipse investigations allowed him to prove some of these notions. His observations were a test of what may be the first scientific hypothesis: A solar eclipse is not the death of the Sun, nor a form of divine punishment. It happens because the Moon slips between the Sun and Earth. Likewise, a lunar eclipse is not a Moon washed in blood or a crimson portent of war. It happens because Earth gets in the way of the Sun, blocking its light from directly reaching the Moon.

Anaxagoras's correct ideas—and his most substantial incorrect one—have not received the credit they are owed for establishing the bedrock of the Western tradition. His works were lost in antiquity, and only transmitted to us through other luminaries of philosophy and science, especially Claudius Ptolemy in his opus the *Almagest*.* As a result, Anaxagoras has been unrecognized and unappreciated—just like the Moon's role in originating Western scientific tradition.

For millennia, credit for the conception of modern science has gone to Aristotle, along with Plato, Aristotle's teacher and the most famous student of Socrates. The "pre-Socratics" are often dismissed as pre-scientific thinkers with odd ideas. This is not their fault. Despite all their creativity, they could not offer any proof for their ideas.[2] The pre-scientific era had not equipped them with any way to test or demon-strate their theories, truthful as they might have been. In his dialogue on the soul, the *Phaedo,* Plato describes Socrates's fascination with Anax-agoras. His revelations were astonishing for a youthful Socrates, who enjoyed learning facts about the shape and location of Earth, as well as speculation about the Moon. But Socrates was disappointed that Anax-agoras didn't lay out any reasons why things were the way they were. "I thought . . . he would go on and explain what is best for each and what is good for all in common," Socrates complained. "My glorious hope, my friend, was quickly snatched away from me. As I went on with my read-ing I saw that the man made no use of intelligence, and did not assign any real causes for the ordering of things."[3]

The tradition of pooh-poohing people before Socrates is actually as old as Socrates. But the revelations of the pre-Socratics, and especially Anaxagoras, were unique in the history of science to that point. Anax-agoras realized things that no one had realized before, or even thought to assume; his ideas upended millennia of human thought and helped to germinate a new form of thinking about nature.

Through observers like Anaxagoras, the Moon became a tool for un-derstanding the universe. By the middle of the first millennium B.C.E., the study of celestial bodies was evolving into a form of study for its own sake. The patterns in the sky had transcended their utility as tools of timekeeping or of augury. By the time of Anaxagoras, the motions of the heavens, and the calculations of those motions, were not just about as-trology. Instead, Greek scholars sought universal truths. They began to move from myth to logos—away from supernatural explanations con-

---

* We will get to Ptolemy, and his tortuous, vexatious epicycles, soon enough.

jured by the human imagination and toward rational thought and observation of natural phenomena. Accumulating knowledge for the
maintenance of power was less important than accumulating knowledge to make sense of the world. Anaxagoras's observations marked the
first time people began to do this in earnest, and he was, like many after
him, focused primarily on the Moon.

○ •

ANAXAGORAS WAS FROM Ionia, across the Aegean Sea from Greece in
what is now Turkey, and arrived in Athens in 480 B.C.E. during the
Greco-Persian Wars. He came from the great Persian tradition of skywatching and brought a distinctly Persian spirit of scientific inquiry to
his adopted city, where he probably traveled as a war veteran or refugee.
It was good timing. Athens was in what would later be considered its
golden age, and its thinkers had just begun to study nature for its own
sake, and to study thought itself. They were also in the process of formu-
lating democracy—rule of the people, by the people—a dramatic change
from the Babylonian or Chinese ideas of kingship as a divine right.

Like most serious philosophers of his day, Anaxagoras was fixated on
astronomy and all the goings-on in the strange realm above the world.

Between 550 and 450 B.C.E., philosophers were circulating a few unusual new ideas about that realm, which young Anaxagoras himself
would have heard. One was from a man named Thales of Miletus. Many
historians, dating back to the early Greeks, report that Thales was the
first to successfully predict a solar eclipse, which occurred as he said it
would in 585 B.C.E. But we don't know how he predicted it. Thales didn't
leave any written theory or ideas about how to arrive at such a precise
calculation. Later philosophers, especially Aristotle, approached Thales
with skepticism. Thales might have gotten lucky, or he might have somehow studied Babylonian eclipse records with great care; more likely, he
paid attention and picked out a pattern, noticing that solar and lunar
eclipses come in pairs, usually about two weeks apart. Anaxagoras
would likely have known about Thales's prediction and the repeating
eclipse pattern.

Another radical idea circulating at the time held that the Moon is not
illuminated on its own, but reflects the light of the Sun. "A light by night,

wandering around earth with borrowed light/ever gazing toward the rays of the Sun," reads a poem by the philosopher Parmenides.[4]

You can try it for yourself. Go watch the Moon, starting with the next new phase, and you will have a hard time *not* seeing it this way.*

We're not sure who thought of this first, Anaxagoras or Parmenides, but, ultimately, it doesn't matter, because Anaxagoras's ideas had more staying power. He was the first person to synthesize the strange, multifarious ideas about the cosmos—some from the great civilizations in Mesopotamia, and some in Greece—and come up with a viable new version of reality. Anaxagoras was the first to offer a geometrical, scientific approach to astronomy, and his ideas form the bedrock of all empirically minded astronomers to follow, from Ptolemy to Copernicus.

"The Sun imparts to the Moon its brightness," Anaxagoras wrote. Equipped with this hypothesis, he watched the eclipse of 478 B.C.E. with alacrity. If the slivered crescent Moon appeared very near the Sun, as it looks just before or just following the new Moon, then where did the Moon go during a solar eclipse? Probably, he reasoned, right in front of the Sun. This meant a heavenly body was blocking another heavenly body.

Anaxagoras had no instruments, no math, no data to draw these conclusions. All he had was his ability to ask questions and a top-notch brain to work out whether he was right. His idea that eclipses result from a blockage of Sun or Moon was a truly remarkable leap from the many weird ideas of the day. The early philosopher Anaximander, for instance, said the Moon was a vast ring and that humans viewed it through a small portal. Anaximenes said it was a flat circle, like a leaf.

---

* Start out when the Moon is new. Find the crescent hanging low in the sky at dusk, before it's dark outside. The Moon will follow the Sun down, sinking on the western horizon before night fully falls. The next night, look again. The Moon is thicker now, and a little higher in the sky when you first spot it. Keep watching, and within a few days, the Moon is half illuminated—a pie sliced in two, with the visible side facing the early-evening Sun. The Moon is full when the Sun is setting, and in the following days, the Moon shrinks again. By last quarter, you can see it just ahead of the Sun in the early-morning sky, once again with its luminous half facing our nearest star. The Sun lights the Moon. It's obvious, once you see the pattern often enough. And once you make this conjecture, that the Sun is what lights the Moon, your repeated observations turn into something else. They turn into evidence.

Xenophanes said it was a cloud. Heraclitus said it was a bowl full of fire, which faced us head-on once a month when the Moon was full.

Anaxagoras alone said the Moon was "Earthy" and had plains, mountains, and valleys. This was an astounding leap from centuries of tradition. His initial insights about the Moon were empirical, not merely philosophical musings on a hypothesis. His thoughts and ideas were grounded in his own observations, and by continually observing, he accumulated evidence that bolstered his ideas. Anaxagoras was building the outlines of a scientific theory for the first time, and in a way that no Greek before him had done.

Eventually, evidence for Anaxagoras's theories expanded beyond the realm of the Moon, to the rest of the celestial sphere. Twelve years after the eclipse, searing hot evidence from above came crashing down to Earth to show that his ideas about the heavens had merit.

In late spring 466 B.C.E., a comet appeared in the early-morning sky, its tail striping the entire dome of the heavens. It grew brighter every day as it fell toward the Sun, and falling stars seemed to streak across the sky at night. On July 18, 466 B.C.E., the comet disappeared in the Sun and reappeared in the evening sky, as the Sun set. Soon afterward, a bright fireball lit up the daytime skies of Ionia. A huge meteorite thundered to the ground, and a burned rock the size of a wagon landed near the town of Aegospotami (now Gallipoli).

It immediately became a tourist attraction, and for good reason. A wagon-sized meteorite would have come from a positively enormous space rock. It would have produced an earsplitting sonic boom and left a sizable crater. Reports of the spectacular fall from heaven are peppered throughout antiquity, from Aristotle himself to Pliny the Elder and Plutarch of Chaeronea, a seminal Roman historian and biographer.

"It is shown to this day by the inhabitants of the peninsula, who stand in awe of it," Plutarch writes in *Lysander*.[5] "It is said that Anaxagoras predicted that one of the bodies entangled in the heaven might, if there were some slip or agitation, break off and fall or be cast down."

The claim that Anaxagoras predicted this particular meteorite fall is even more spurious than the claim about Thales and his eclipses, because we have no record of how he could have done it. But Anaxagoras did realize something new, and for this he deserves credit. He may have predicted "the *possibility* that a rocky body could fall from the sky given the nature of heavenly bodies he hypothesized," said Brigham Young University philosopher Daniel W. Graham.

"Immediately the story was, 'Well, Anaxagoras predicted that,' because his was the only theory that had heavy bodies in orbit," Graham told me. "There is a line in Homer about Zeus throwing rocks at the Earth, so before that time, the explanation would have been, 'Zeus must be angry at someone.' But instead, people said, 'Anaxagoras is right.' They didn't see this as a divine portent anymore. This was a prediction of an event that could have happened, and when it did, it was taken as a confirmation of his theory."

If the rock that fell on Aegospotami came from above, from the realm of the Sun, Moon, and stars, then those objects were probably made of similar stuff as the rock, Anaxagoras reckoned. He extended this idea to everything else in the sky. He argued that the Sun was a hot mess of fiery metal. He said the stars and planets were giant pieces of stone that had been shorn from Earth and later ignited. He said we don't feel heat from the stars because they are so far away. He came up with explanations for the light of the Milky Way and the solstices. Anaxagoras in his own time earned the nickname "Mr. Mind," according to the third century C.E. historian Diogenes Laërtius,[6] because he believed the cosmos is controlled by *Nous,* a great mind or intelligence that served as a cosmic architect. Socrates, the great philosopher of Athens, liked this notion, too. Anaxagoras believed that the universe's original state was a mixture of all its current components, but not evenly distributed, and that *Nous* organized it all. He argued that everything that exists contains some of this primordial material—anticipating, in a way, the basic outlines of what we now call the Big Bang.

Some of these ideas would be dismissed in antiquity only to be proven right millennia later. But the meteorite, above all, showed that Anaxagoras correctly understood a key feature of the skies above. The celestial bodies are rocks.

Once people got over their fear of the Gallipoli meteor, they approached the steaming celestial visitor and realized what it was. The thing had clearly fallen from above, like Anaxagoras said it might; they knew it was true because so many people saw it fall. They knew the rock was real because they could touch it. It was a tourist attraction for the next five centuries. The meteor was Earthy; it was a thing with mass that could be perceived; it was not made of the ether or some cloud. And it came from the realm overhead, somewhere above Earth and closer to the Moon. (There's a reason we still call the study of Earth's weather, its atmospheric phenomena, *meteorology.*)

The comet was visible for as many as eighty days, according to reconstructions of the event using modern astronomy software. Its tail was gigantic, spanning almost the entire sky. We know of one other comet that has done this, and done so repeatedly, every seventy-six years. The comet[*] could have been Halley's, named for a pal of Isaac Newton's who predicted it would reappear, and became famous when it did, long after his own death. Anaxagoras may have been the first to record it, and to glean some new insight from it. Maybe we should call it Anaxagoras's Comet.

ALL OF ANAXAGORAS'S unorthodox ideas refuted earlier theories. There were no giant cosmic bowls up above, there were no flat leaves—there were rocks, and the Moon was one of them. The notion that the Sun illuminated the Moon meant the phases of the Moon made sense like never before, and so did the nature of eclipses. Between Anaxagoras's experiences of the eclipse and the meteor, his ideas about the heavens began not only to make sense, but to suggest evidence that could be tested. Anaxagoras represents a bridge between philosophy and the history of science: a man of lofty ideas who realized he could observe the world, and through observing, test those ideas.

But there was a problem. During Anaxagoras's life, these ideas were considered not only ridiculous, but heretical. The Sun was the god Helios, who drove his chariot up and down the sky at sunrise and sunset, while "bright beams from his person blaze dazzling."[7] At night, his horses rested under Earth. The Moon was the goddess Selene, Helios's

---

[*] There's no way to prove whether Anaxagoras's meteorite came from the comet—it might have been a coincidence—but comets do cause meteor showers and can cause meteorite impacts. Though the vast majority of meteorites that land on Earth are cleaved from asteroids, some researchers believe a few of Earth's largest impacts have come from comets, in part because their orbits around the Sun grant them high speeds when they get close to us.

sister, whose long-maned steeds only reached full speed in the middle of the month. In Hesiod's *Theogony,* 225 years before Anaxagoras, meteorites are attributed to "The Hundred-Handed Ones," giants with one hundred arms and fifty heads who hurled great boulders to help Zeus defeat the Titans on Mount Olympus. To argue that the Sun and Moon were merely hot rocks in the sky was impious, to say the least.

So it should not be surprising that Anaxagoras was eventually put on trial in Athens. Even his friend Pericles—the heroic general, builder of the Parthenon, and founder of Greek democracy—could not protect him. Anaxagoras was convicted of heresy and sentenced to death.

He fled home to Ionia, arriving in a city called Lampascus. But he would not be silenced. His book was still published in Athens, and anyone could buy it on the street for a drachma, according to Socrates. In his infamous trial, that great philosopher was accused of sharing Anaxagoras's blasphemous beliefs. Plato's *Apology* tells the story of how Socrates was accused of "corrupting the youth" and of disbelieving in the gods of Athens. Socrates defends himself by parodying his accusers.[8]

"Do you mean that I believe neither the Sun nor the Moon to be gods, like other men?" Socrates says at one point, exasperated. The judge Meletus retorts that yes, you think the Sun is a stone and the Moon is an Earth. Socrates is annoyed: "My dear Meletus, do you think that you are prosecuting Anaxagoras? You must have a very poor opinion of these men, and think them illiterate, if you imagine that they do not know that the works of Anaxagoras of Clazomenae are full of these doctrines."*

Socrates says the book was widely available, although no copies survived from antiquity. A long list of Greeks built on Anaxagoras's work in their own ways, and their works survive, so those scholars are more well known than Anaxagoras now.

PLATO, BORN AROUND 428 B.C.E., the year Anaxagoras died, wrote about Anaxagoras's ideas but he did not share in them. He was suspi-

---

* Socrates put together an impressive self-defense, and at one point his judges offered clemency against the death penalty if he would admit to minor charges of impiety and corruption. He refused, and poisoned himself by drinking hemlock tea. He died in 399 B.C.E.

cious of astronomy because he disliked any conclusion that was based on observation; he thought we should study the mathematical motions of the heavenly bodies, not the motions they appear to make as seen from our limited perspective. He taught that reason can yield truth, but only pure reason that was unmarred by experience.

Plato also gives us the idea of a supreme being. In his dialogue *Timaeus,* God is a creator who deliberately crafts reality and creates time. Philosophy, the study of knowledge and existence, flowed from this mathematical construct. Plato was explaining something that the Scots of Warren Field had figured out eight millennia earlier, albeit in a much more literal way. He understood that the Moon and the Sun give us a concrete representation of the passage of time. But to Plato, time was more than a means of reckoning the past and figuring the future. For Plato, time was the originator of philosophy; "number was the explanation of the world," as the historian of philosophy Bertrand Russell explains it. The cosmos owes its nature to the numbers that compose its very essence—numbers that Pythagoras and his students had revealed.

"God made the Sun so that animals could learn arithmetic," Russell writes. "The sight of day and night, months and years, has created knowledge of number and given us the conception of time, and hence came philosophy."[9]

Though he knew the Moon was crucial for timekeeping—it was the basis of the Greek calendar, after all—Plato was more obsessed with ideal forms, the nature and movement of the soul, and the concept of Mind, *Nous,* than he was with the Moon. His most famous student, Aristotle, saw things differently. Aristotle saw beauty in the order of things, but more than Plato, he wanted to know why things were the way they were. He wanted to understand natural processes, not just note that they occurred. This is why Aristotle, not Anaxagoras, gets credit for ushering in what we'd consider the pursuit of science. Science, according to Aristotle, is observation and the unraveling of mystery in the pursuit of truth.

In Aristotle's cosmology, *De Caelo,* the heavens originated at the Moon, which he considered the boundary of the realm of perfection. To Aristotle, perfection did not mean the divine or a deity, the way Plato or religious people thought about it. Perfection to Aristotle meant the perfect ideal of something. Everything below the Moon was in the realm of change, corruption, and imperfection. The Moon and beyond were perfect. He explained away the Moon's mottled appearance by arguing that it was contaminated by Earth.

The most important contribution of Aristotle's Moon-centric cosmology was the idea of astronomy as a scientific discipline. Anaxagoras took the first small steps, but after Aristotle, astronomy was more science than art. It was practiced in order to figure out the universe, purely for the sake of understanding. Studying the Moon was the honorable work of high-minded rational gentlemen, not cultish sky priests tasked with hedging bets for twitchy sovereigns. Thanks to the Moon, people started imagining the skies as a classroom, where lessons could be learned about the nature of reality. The educated were to observe the cosmos, to explain their observations in theory, and to check those theories against other observations.

Anaxagoras did this first—arguably, the Babylonians under Nabonidus began a version of it even earlier—but Aristotle was the one who made investigation his explicit mission. He founded his own school, called the Lyceum, a version of the temple academies of the past and Plato's own Academy. Students were surrounded by maps, scrolls, and books, the first library in the ancient world. They had laboratory tables for dissecting animals and large lecture halls where students could hear Aristotle hold court. His successors carried the ideals of the Lyceum into the future and ensured they would reach all the way to the Enlightenment. The American historian Arthur Herman, in his Plato-vs.-Aristotle biography *The Cave and the Light,* notes that Aristotle saw his philosophy, including his Moon-centric cosmology, as a complete picture of reality. It encapsulated everything. But "by stressing the power of observation as the main source of knowledge, he had let the genie run free," Herman writes. Others would carry this scientific method forward, make their own observations, and ultimately surpass Aristotle's neatly packaged cosmos. The concept of seeking astronomical knowledge for its own sake would power the minds of the most consequential scientists to follow, twenty centuries later: Nicolaus Copernicus, Johannes Kepler, and Galileo Galilei.

But not yet.

To FOLLOW IN the footsteps of Anaxagoras and Aristotle, the makers of the Western scientific tradition first had to shake off these thinkers' most egregious mistakes. Both Anaxagoras and Aristotle believed, mistakenly, that the Sun revolved around Earth. They probably believed this because of the Moon; it makes a lot of sense, especially if you're Mr. Mind and look to the Moon to figure out reality.

At least one ancient Greek figured out the truth, but his works were lost until after Copernicus independently made this realization a second time. Aristarchus of Samos (310–230 B.C.E.) realized that when the Moon is exactly half full, the angle between Earth, Moon, and Sun will be ninety degrees. He calculated the ratio of the Earth-Moon distance to the Earth-Sun distance and found out the Sun is much more distant than the Moon, though it appears to be the same size in the sky. This revelation led to an absolutely astonishing conclusion: The Sun is much bigger than the Moon. The monarch of the night is therefore not on par with the ruler of the day. Aristarchus realized that given these huge sizes and distances, Earth must revolve around the Sun, and not the other way around. But nobody listened to him.

Despite these geocentric blunders, Greek astronomers after Anaxagoras learned a great deal about the motions of the heavens. And they began to realize they could use their knowledge of those heavenly movements to seize control of time. Just like the Scots of Warren Field and the Únětice warlords who hammered the Nebra sky disk, the classical Greeks knew that time is a cudgel.

Writing around the same time as Anaxagoras, Meton of Athens figured out that nineteen solar years and 235 lunar months are almost the same length: the Metonic cycle. He introduced this cycle into the Attic calendar, which was based on the cycles of the Moon. Then in 331, Alexander the Great, a student of Aristotle's, became the next to conquer Babylon, that great city on the Euphrates. He ordered the Babylonians' astronomical tablets translated into Greek. The Greeks found a wealth of knowledge within, and Alexander became another in a long line of rulers to grasp the immense power that derives from being in charge of time. He who controls the calendar controls society, as the builders of Warren Field knew, as the makers of the Nebra sky disk knew, as the first farmers of Mesopotamia knew. Time confers power to whoever

commands it, and the Moon remained the simplest and most predict-
able way to seize that command.

Working on behalf of Alexander, a fellow student of Aristotle's named
Callippus used the Babylonian records and the knowledge of his Greek
predecessors to once again recalculate the length of the lunar month.
Callippus proposed a new calendar, which started on June 28, 330 B.C.E.,
eight months after Alexander captured Babylon.

The new calendar, which all later Greek astronomers used, runs for
seventy-six years, or four Metonic cycles. After 940 Moon months, at the
end of every fourth nineteen-year period, Callippus's calendar drops a
day. This keeps the lunisolar calendar even better aligned with the seasons of the solar year.

Building on the religiously motivated observations of the Babylonians and the scientific reasoning of the Greeks, humans in antiquity
got better and better at making calendars. As a result of their increasingly precise timekeeping, the Moon would fall, for the first time, from
the height of its influence over the human mind.

◯ ˙

ALEXANDER THE GREAT had conquered the world, in a manner of
speaking, by the young age of twenty-five. He controlled a vast empire
that stretched from the Adriatic Sea to the northeastern edge of India.
Long after his empire collapsed, long after Alexander's death in Babylon, a thirty-nine-year-old Roman nobleman named Julius Caesar was
vexed that he had not yet done the same. But he ultimately would. And
his mastery over the calendar, especially as it related to the Moon,
would quickly become even more consequential than Alexander's.

By 61 B.C.E., Julius Caesar was a well-liked general with outsized political ambitions. He was a self-obsessed looker, a lover to countless
women and probably at least one man, an appreciator of fine art and
jewels, and a skilled military commander. After various heroic military
exploits, he was appointed governor of Spain, and brought it under
Roman control. It was not enough for him, however. Gaius Suetonius, a
biographer of the Caesar dynasty, says Julius saw a statue of Alexander
the Great during his Spanish adventures and "was overheard to sigh im-

patiently," frustrated that he was still a mere general.[10] Caesar vowed to make his own mark. So in 59 B.C.E., he returned to Rome to run for consul, basically an elected co-executive over the Senate. He needed allies, and fortunately for him, there were other ambitious, petty, and arrogant military men to connect with. Caesar struck up an alliance with two: Gnaeus Pompeius Magnus, commonly known as Pompey, and Marcus Licinius Crassus. They formed a triple pact, jointly swearing to oppose all legislation any one of them disliked. The First Triumvirate was a fragile partnership, because each mistrusted the others.

The downfall of the First Triumvirate, their imbroglio, and the rise of Caesar as emperor is one of the best-known stories from antiquity. We can mostly thank William Shakespeare for that, though Caesar's exploits were renowned hundreds and hundreds of years before any English writer penned a play about him. There is something among all his adventures that neither Caesar nor Shakespeare gave much thought to, however. The Moon was central to three important events in Caesar's life—including its end.

A DECADE BEFORE Caesar ran for consul, Crassus was failing in his effort to crush the enslaved gladiator rebellion led by Spartacus—the man played by Kirk Douglas in the 1960 movie—and the Senate called in General Pompey as backup. Crassus never got over this slight. Pompey, in turn, was annoyed by Crassus's reputation as the wealthiest man in Rome, and by Crassus's attempts to needle him and block him from spending state money on bribes. Caesar intervened as a truce maker and solidified his own partnership with Pompey by arranging a marriage between Pompey and Caesar's daughter, Julia.

While Pompey and Crassus were co-consuls, Caesar had full support to rampage through Gaul and Britain, racking up military victories and glory for Rome. He may have slaughtered as many as a million people[11] in the whole region, though historians like Pliny the Elder later struggled to determine the exact number, instead accusing him generally of "a crime against humanity."

Then three key events happened between 55 and 53 B.C.E. that would change everything.

On August 23, 55 B.C.E.,* Caesar attempted to invade the southern coast of England, claiming that the Britons had helped his enemies in Gaul. (Suetonius said he was looking for pearls.) Archaeologists found evidence in 2017 C.E.[12] that his forces landed in Pegwell Bay, fifteen miles north of Dover's white cliffs, where Caesar set up a camp to prepare for the rest of the legions to join him. But the cavalry was not coming. The Moon ensured that.

Caesar, who was used to the mild waters of the Mediterranean, was unprepared for the high tide in the English Channel. His navigators weren't ready. It was a full Moon high tide, one of the two highest tides of the month—which the Allies exploited two millennia later, on D-Day. Caesar's ships started taking on water. His troop transports were dashed against one another. Some of his ships wrecked and others were so damaged they couldn't make it home. He turned tail and fled back to Rome, with neither conquest nor pearl.

But Caesar had nonetheless done something new. He had planted the flag of Rome on the soil of a new world. At the time, some Romans weren't even sure Britain really existed. It was a thousand miles away, across the Alps and multiple rivers and the English Channel. Caesar and his troops proved that it was real, and they would be back to take it over.†

In Caesar's own telling, his standard-bearer from the Tenth Legion jumped into the high waters to fight the Britons. "Jump down, soldiers," he shouted,[13] "unless you want to betray our Eagle to the enemy—I at least shall have done my duty to the Republic and to my commander."‡

The citizens of Rome were so impressed to hear of their Eagle banner on British soil that the Senate ordered a twenty-day festival in thanksgiving, which included gladiatorial games, feasts, and other activities

---

* This date is according to the Julian calendar for 55 B.C.E. Caesar would not inaugurate that calendar for another decade, but the dates of his exploits as written here were "translated," in antiquity, to the timekeeping system he made up.

† Julius Caesar never gave up, so he tried again the following year and succeeded.

‡ That Eagle symbol would last through the centuries. It was adopted by, among others, the Nazi Party during World War II and the United States, where it became the insignia of the Apollo 11 mission and the name of the human-inhabited spaceship that touched the Moon for the first time.

that endeared the people to their swashbuckling general. Crassus, Pompey, and others watched it all with dismay. Crassus, still smarting from his unsuccessful battles against Spartacus, also wanted to earn military glory. He figured a war against the Parthian Empire, which controlled much of Mesopotamia and what is now Iran and Turkey, would be "simple, glorious, and profitable," in the words of the Roman historian Appian.[14] Crassus decided to declare war. The Senate tried to block him, because the Parthians had done Rome no harm, but Crassus went anyway. In spring 53 B.C.E., he marched his army through the Mesopotamian desert. On June 6, under a first-quarter Moon, they were met by the forces of the Parthian general Surena. The Parthians decimated the parched Roman troops and Crassus was beheaded.

Crassus died along with thousands of his men in a city the Romans called Carrhae, in what is now Turkey. Carrhae had another name: Harran. Crassus died seeking glory in the Moon City of the East. The city of Nabonidus's birth, the city where his mother, Adad-guppi, was High Priestess of Sin, the city where Nabonidus restored the Moon God's temple in defiance of the priests of Babylon. Crassus never saw Rome again.

EVERYTHING FELL APART after the Battle of Carrhae. Crassus was dead and so was the First Triumvirate. Caesar's only child, Julia, had died in childbirth a few months before, and so had her infant. Caesar's alliance with Pompey was on the rocks. After Crassus was killed, a young bureaucrat named Gaius Cassius Longinus acquired the Roman watch over Syria. Caesar had been in Gaul for nine years, assembling and leading a military force numbering forty thousand troops, and Pompey and the Senate decided that was long enough. Caesar was called home and told to disband his army. He was infuriated, so sometime around January 10, 49 B.C.E., nearly a full Moon, he made a fateful decision. Standing on the banks of the Rubicon, the river that formed the boundary between the wild lands of Gaul and civilized Roman Italy, he mulled over his options. Suetonius says that Caesar and his men were visited by an apparition, who seized a Roman trumpet, ran across the river, and blew the instrument in an obvious omen from the gods. Appian says that Caesar reflected thoughtfully, remarking, "If I refrain from this crossing, my friends, it will be the beginning of misfortune for me; but if I cross, it will be the beginning for all mankind."[15] Suetonius, Appian, and Plutarch all say that he crossed the river defiantly and, quoting an Athenian playwright, said, "*Alea iacta est.*" The die is cast.

It was an act of civil war. Caesar was marching on his own city with his own army. Pompey fled to Alexandria, where he hoped to hide and organize his troops. But the pharaoh of Egypt, Ptolemy XIII, had Pompey killed. Caesar declared war on Egypt, too, killed Ptolemy, and installed Cleopatra as queen. The infighting in Rome now encompassed the entire Western world. Caesar was made dictator in Rome. He held parades and festivals, made his friend Mark Antony second-in-command, paid lavish bribes to politicians to legitimize his takeover, and generally brought about the end of the Roman Republic. He had everything at his disposal. And one of the first things he did was remake the calendar, the ultimate power play, seizing command of the order of time.

DURING THE CONSULAR ERA in Republican Rome, the year was only 355 days long and was based entirely on the cycles of the Moon. The first day of a month, the Calends, from which we get the word "calendar," took place when the new Moon was first sighted. The Nones marked when there would be nine days until the Ides. The Ides was the day of the appearance of the full Moon.

But as we know, the lunar year is not the same length as the solar year. To make the calendar match the real seasons, Roman priests had the solemn duty of occasionally adding an intercalary month to make up for that roughly ten-day lunar lag. This would ensure that the Moon-directed civic calendar kept in time with seasonal holidays, harvests, and festivals, just like the Nebra sky disk and the Babylonian calendar accomplished.

But the Roman calendar keepers did not always add the intercalary month correctly or on time. The Roman Republic was big and messy. Sometimes, people in the far-flung provinces would not hear about an intercalation for months after the fact. Imagine showing up for a holiday dinner after an arduous journey, on horseback if you were lucky, and being told you missed it because it happened six weeks ago. Sometimes the calendar makers were distracted by other affairs of state; often, the authorities would miss an intercalation during wartime. As a result, the calendar year and the natural year were often weeks or months apart. The harvest festival might fall when crops were still growing, and the spring months would fall during midwinter. The calendar mishap happened during Caesar's civil war, so he set out to fix it.

Caesar likely had heard of a unified Moon-Sun calendar, called a lunisolar calendar, while consorting with Cleopatra. The traditional Egyptian calendar had 365 days, comprising twelve months of thirty days apiece, plus an extra five intercalary days. To keep in time with the solar year, the Egyptian calendar would include a sixth intercalary day, a leap day, every four years. The Roman statesman Pliny the Elder writes in his *Natural History* that Caesar worked with Sosigenes, an Alexandrian mathematician, to design his own new calendar and produce a new astronomical almanac. The new calendar divided the year into twelve months with thirty or thirty-one days, except February, which has twenty-eight. An intercalary "leap day" is added to February every four years. It's essentially the same calendar we use now; it has received just one update in the two millennia since.

Remember that because of the way Earth wobbles on its axis, a calendar based on the movement of the Sun against the stars does not quite match the time between spring equinoxes over many years. Those days when Earth is not tilted toward or away from the Sun, giving us an almost equal day and night, define the tropical year. The fix of Caesar and Sosigenes was still slightly off: The time from one spring equinox to the next takes 365.2422 days, and Caesar's calendar makers counted 365.25 days. Caesar's calendar drifted against the solar year by one day every 128 years.* By decree, Caesar reorganized the year 46 as a transitional year, lasting a whopping 445 days—convenient for the first year of a new dictatorship. Romans began their year as usual, but knew they would have three extra months tacked on at the end. After this final "year of confusion," the new, far more accurate calendar finally began on January 1, 45 B.C.E.

It was a dramatic improvement for timekeeping, but it was a tragedy for our relationship to the Moon. For the first time in the history of our

---

* Over the centuries, this added up to a lot, and the calendar was out of whack again.

species, the Moon was separated from time. The days slowly slipped out of sync with the Moon's cycles. The Ides were reduced to signify a numerical meaning, the fifteenth day of a month, rather than the mark of the full Moon. The Julian calendar was revised in 1517 and in 1582 C.E. so that Easter would align with the spring equinox. We now call it the Gregorian calendar, but the version Julius Caesar bestowed upon us is still the predominant marker of time in the modern world, still separate from the Moon's cycles.*

Though he fixed the calendar, Caesar, beloved by the Roman people but not its Senate, was increasingly under siege. In early 44 B.C.E., he mustered one hundred thousand troops, intending to march them to the Moon City of Harran/Carrhae and exact revenge for the death of Crassus. But a few days before he was to leave for the East, his friends and enemies surrounded him in the Senate house. Cassius, the official in Syria who was promoted after the death of Crassus, led the conspiracy. They stabbed him twenty-three times on March 15, 44 B.C.E., the Ides of March. The Moon was full.

$$\bigcirc \; \cdot$$

THE DEATH OF Caesar ushered in another few years of chaos, as his friends and allies fought against the conspirators. His nephew Octavian, later known as Caesar Augustus, ascended to power. The year after Julius Caesar died, Augustus named what is now our seventh month of the year for his uncle: July instead of Quintilis. Roman prestige and power spread through the Mediterranean and the Levant, and into mainland Europe. Augustus's successor and stepson, Tiberius Caesar, levied taxes throughout the empire to pay for its increasingly expensive exploits and public displays. Around 33 C.E., the Roman governor in Judea thought he could trick a man who had been telling people not to worship Augustus or Julius Caesar, who by then had both been deified. The governor, Pontius Pilate, thought the philosopher from Bethlehem would also op-

---

* Many cultures still use a lunisolar calendar, notably the Hebrew, Chinese, Buddhist, Hindu, and traditional Korean, Japanese, and other Asian calendars. The Islamic calendar is a lunar calendar. But the global systems of finance, trade, and politics all use the Western timekeeping system.

pose paying taxes to the emperor, and challenged him on it. "Render therefore unto Caesar the things which are Caesar's; and unto God the things that are God's," the man reportedly replied.[16]

The assassination of Julius Caesar started the downfall of Republican Rome. But the first generations of emperors—of Caesars—ushered in a two-hundred-year-long period of prosperity, expansion, and peace called the Pax Romana. In this time of relative tranquility, thinkers in Rome continued the tradition of Anaxagoras and Aristotle, turning skyward for answers to their deepest questions.

Even though the Moon was divorced from the predominant Western method of civil timekeeping, it still occupied a place of high regard, both in the sky and in the minds of philosophers. If you are inclined to think about the nature of things, how the cosmos is arranged, and what it all means, the Moon will always be an interesting subject. It exists in a realm apart from Earth but is clearly inseparable from Earth. Its appearance changes every night, but it always returns. It is safe to gaze at the Moon for long periods, unlike the blazing Sun. People used the Moon for quotidian reasons like calendaring, but humans through the ages also just looked at it, enjoyed it, and wondered about it.

Many philosophers came up with various reasons why it would not fall down on us all, coming up with creative cosmologies that explained its distance from Earth. Explanations ranged from a circular, flat terra firma surrounded by a river, like a moat keeping the Moon from getting too close, to a flat Earth protected by a heavenly top layer, like a cake with frosting. Some believed that the celestial bodies were harmonic perfect spheres that could never come into contact, lest they pollute one another. From the holy patron of Mesopotamia to Aristotle, scholars had also imagined the Moon as a haven for the divine and a harbor for souls. Plato's dialogues suggested that the Moon and Sun were places we might visit once departed, if we had lived a good life. But arguably the most famous discussion of the Moon and the soul, and arguably the most important, comes from the Roman statesman and historian Plutarch of Chaeronea.

Plutarch, who lived between 45 and roughly 119 C.E., is best known for his *Moralia* and *Parallel Lives,* paired biographies of Greek and Roman statesmen and military leaders. He's the main reason we know what we know about Anaxagoras, and one of the reasons we know about Caesar's adventures abroad.

But Plutarch was also a prolific writer of philosophy, mostly in the form of dialogues. His treatise *On the Apparent Face in the Orb of the Moon*, usually abbreviated as *De Facie*, attempts to describe the nature of the Moon and its relationship to the nature and fate of the soul. It's practically a compendium of early astronomy, covering Plato, Aristarchus, Anaxagoras, and more. The mathematician and astronomer Johannes Kepler, who was keenly interested in the Moon, produced his own translation of *De Facie* and considered publishing it. He called it "the most valuable discussion of the Earth's satellite to come down to us from antiquity."[17]

De Facie is a complicated dialogue among several young men who are out for a stroll. As the characters promenade through the city, probably Rome, they chat about the nature of the Man in the Moon.* One character argues the Moon's apparent face is just a reflection of Earth's oceans. (Seventeen centuries later, astronomers still thought this was the case.) Some think the Moon is a "mixture of air and gentle fire," while others say it's "a body of weight and solidity" like Earth. The characters tease one another for their unorthodox ideas, discussing geometry, attempting to deduce the Moon's size, and trying to determine why it does not fall down on Earth. They compare its features to the Caspian Sea and the Pillars of Hercules (the rocky outcrops that flank the Strait of Gibraltar). They even mention the bizarre eclipse rituals of Babylon: "Most people have the custom of beating brasses during eclipses and of raising a din and clatter," Plutarch writes.

Then a character named Theon blurts out a remarkable notion.

"I should like before that to hear about the beings that are said to dwell on the Moon—not whether any really do inhabit it but whether habitation there is possible. If it is not possible, the assertion that the Moon is an Earth is itself absurd, for she would then appear to have

---

* In Western folklore, as in Plutarch's essay, the full Moon appears to have a face, depending on how you look at it. Other cultures imagine a rabbit with large ears. If you're looking for the apparent face, the large, dark splotch that forms the "right eye" is Mare Imbrium, the Sea of Rains. This is where geologist/astronaut Jack Schmitt raked up lunar soil in 1972, lifting troctolite 76535 off the Moon and into history.

come into existence vainly and to no purpose."[18] Someone replies that sure, the Moon is probably inhabited, but by an ethereal type of being who views our world as slimy and gross and the Moon as the only true paradise.

This whole discussion is novel in written history. If people had posed these questions before, no one had written them down, at least not in a manuscript that survived the ages or the keen eye of the pious. What is the meaning of the Moon? What is the point of the celestial bodies? *Are we alone?*

Humans were beginning to understand the Moon as a world unto itself. They could imagine the Moon as a place with features that would look familiar. Far from a mysterious, distant orb, the Moon was just like home. This realization further diminished the Moon's stature for us. As we attempted to make sense of the Moon, it necessarily grew less mysterious. It had already fallen from its secular, civic height in keeping time; now, through the unromantic lens of philosophy, the Moon had become less holy, less spectral, and less vital.

Plutarch's characters ultimately established that the Moon was "a second Earth," some version of a terrestrial body. They agreed with Anaxagoras that the Moon was Earthy. They even imagined other beings living on it. *De Facie* provided the seed, and, sixteen centuries later, Kepler cultivated what became the next seminal shift in the Moon's history.

BY THE TURN of the first millennium C.E., humans had realized that the Moon was closer to Earth than the Sun is. It was no longer of equal stature to the Sun, but a distant second. We had realized it was not a god, but a mere object, and a plain one at that: not a star, not a heaven, just a rocky, spotted ball like Earth. The Moon wasn't even the Western world's timekeeper anymore, just an interloper in the solar-focused year. And philosophers had begun to argue that it was neither perfect nor divine, but merely a place. As Western civilization entered the age of Christianity, the Moon's importance was on the wane. And it remained in this diminished state for more than a millennium, until the scientific revolutionaries of the seventeenth century hurled it back to the forefront.

PART III

HOW WE MADE
THE MOON

CHAPTER NINE

The Moon in Our Eyes

My dear Kepler, what would you say of the learned here, who, replete
with the pertinacity of the asp, have steadfastly refused to cast a
glance through the telescope? What shall we make of all this?
Shall we laugh, or shall we cry?

—Galileo Galilei, *Frammenti e lettere*[1]

PEOPLE THROUGH THE ages used the Moon for timekeeping, for accumulating power, and for understanding the world. But by the sixteenth century C.E., the Moon mainly caused cosmological misunderstandings instead.

The societies of Mesopotamia were content to measure the motions of the celestial bodies without trying to answer why things move the way they do. They wanted to know the heavens for religious reasons. In their careful studies, they learned a form of science, but the Moon was the fly in the ointment. If we had no Moon going around Earth, maybe the Babylonians would have understood things more accurately. They may have figured out that when Venus appears as both morning and evening star, it's because Venus circles around the Sun. The weirdly backward motion of Mars, Jupiter, and Saturn would also be explained by their mutual orbits of the Sun. People may have grasped the truth that Earth, closer to the Sun with a shorter orbital period, just overtakes the other planets sometimes, like a fast runner lapping a slower one. If the Moon did not so obviously revolve around Earth, there may never have been a reason for humans to believe that everything else did, too.

The classical Greeks were more intentional about studying cosmic reality, but they were curtailed by their limited knowledge and their obsession with how the universe was controlled—which many, from Anaxagoras to Aristotle, thought was via *Nous,* some indistinct form of consciousness. Aristotle did come up with explanations for the nature of the universe, making observations and drawing conjectures in his *De Caelo.* But many of his conclusions were way off. He wrote, self-assuredly but incorrectly, that the Sun circles Earth just the way the Moon clearly does. Early thinkers fixated on this fact—the Moon goes around Earth—and extrapolated from there: so must everything else.

This arrangement is certainly logical, to be fair. There is no obvious reason why different celestial orbs should do different things. Especially in the early Christian world, when Scripture was considered infallible and the only mechanism for understanding reality, why would anyone assume the Sun and Moon do not behave the same way? Moreover, why would such a perfect system, endowed by a perfect creator, have more than one means of heavenly motion? The geocentric model was easy to accept because it made sense. As long as you didn't ask too many questions.

By the turn of the seventeenth century C.E., people did start asking those questions, or at least they finally began writing them down and publishing them. Men like Thomas Harriot, Johannes Kepler, and Galileo Galilei changed the way we saw the cosmos. They were the first modern scientists, and they bequeathed to us a new era and a new way of thinking. Confronted with unfamiliar lands beyond the seas, the scientific thinkers of this generation realized the received wisdom of antiquity—and the Church—was not the last word on the world. Confronted with new realms on display in the heavens, they realized knowledge could be sought, and then attained. These men promoted and then proved the revolutionary heliocentric theory of Nicolaus Copernicus. They began to figure out, for the first time, how the heavens are actually arranged. Using the height of modern technology, the telescope, they were finally able to learn about the cosmos using more than geometry and conjecture, the way Anaxagoras and Aristarchus and Plutarch had learned. These men could see like no one before them, and the first place they looked was the Moon. But they did not merely see the universe more closely. They saw it more truthfully. In the seventeenth century, people finally saw the heavens as a tapestry of detail and realism.

Marines wounded in the landing on Tarawa are towed out to larger craft bound for base hospitals in the South Pacific. The Battle of Tarawa, in November 1943, ended in the worst casualties in United States Marine Corps history, because the tide did not rise as predicted, owing to the Moon's position relative to Earth.

The Moon's far side looms beyond the *Orion* spacecraft in this image taken November 21, 2022, during NASA's Artemis I mission. The darkest spot on the Moon is Mare Orientale. If you squint, the slightest hint of a cheery rose color is visible on the Moon's right limb.   PHOTO BY NASA.

Facing page: Numerous theories have been proposed to explain how the Moon came to be, but none is definitive yet.

Four Ways to Make the Moon

As the leading theory for the moon's formation runs into problems, scientists have floated other ideas for how the moon came to be.

Giant Impact Model

This classic theory, developed in the 1970s, holds that a Mars-size rock called Theia smashed into the young Earth. The impact created a disk of debris that eventually coalesced into the moon. Yet recent studies have revealed a conflict: Computer simulations of the event suggest the moon should be made of mostly Theia-like material, while lunar geochemistry research suggests that the moon is made of Earth-like material.

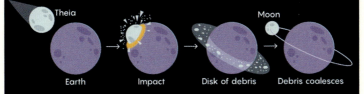

| Earth | Impact | Disk of debris | Debris coalesces |

Synestia

Perhaps Theia struck the proto-Earth with enough energy to vaporize both objects, forming a new cosmic structure called a synestia. This rotating cloud of hot debris could have thoroughly mixed material from Theia and Earth, leading to a Earth-moon system with identical geochemistry.

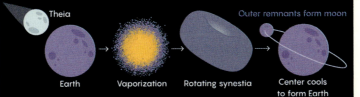

| Earth | Vaporization | Rotating synestia | Center cools to form Earth |

Moonlets

Instead of one giant impact, perhaps many smaller impacts created the moon. In this model, each moon-size impactor creates a debris disk that eventually coalesces into a moonlet. Successive impacts create additional moonlets that all eventually combine to form the moon.

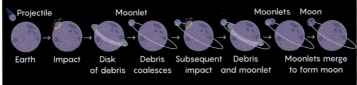

| Earth | Impact | Disk of debris | Debris coalesces | Subsequent impact | Debris and moonlet | Moonlets merge to form moon |

Twin Collision

Perhaps the simplest alternative is that Theia was made of the same kind of material that the young Earth was. This possibility challenges much of what we know about the formation of planetary systems, however.

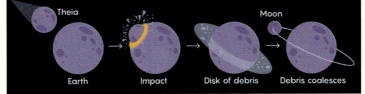

| Earth | Impact | Disk of debris | Debris coalesces |

Hipparchus of Nicaea attempted to map the entire night sky and is credited with figuring out the precession of the equinoxes, in which Earth's axis seems to wobble like a top over time. ILLUSTRATION C. 1880. ANN RONAN PICTURE LIBRARY, LONDON. USED WITH PERMISSION.

The Venus of Laussel, a carving in relief dating to twenty-seven thousand years ago, is thought to represent a connection between human fertility and the lunar cycle. The bison horn in the woman's hand is etched with thirteen notches, which might represent the lunar cycles in one year.

SISSE BRIMBERG/GEO IMAGES COLLECTION, ART RESOURCE, NEW YORK. USED WITH PERMISSION.

Hilary Murray and Charlie Murray stand in Warren Field, which sits atop the ruins of the world's oldest known lunar calendar. It was made from a series of pits in the ground in a field adjacent to a fifteenth-century castle in Aberdeenshire, Scotland. LEFT: PHOTO BY THE AUTHOR; ABOVE: COURTESY OF MURRAY ARCHAEOLOGICAL SERVICES LTD.

The stone circle of Tomnaverie, in Aberdeenshire, Scotland. These monuments are found only in northeastern Scotland and are characterized by a large stone on its side, flanked by two upright stones, usually on the south or southwest arc of the circle. Tomnaverie's sideways stone is in the center and may be aligned to match the cycles of the Moon. PHOTO BY THE AUTHOR.

The Nebra sky disk, which dates to sixteen hundred years ago, is one of the most important archaeological finds from the European Bronze Age. The circle of hammered gold might be the Sun or an eclipsed Moon, facing a golden crescent Moon that is a few days into the lunar cycle. The Pleiades constellation, also known as "the Seven Sisters," is visible. The golden arc at the bottom may symbolize the Sun's apparent path across the sky between solstices.

The Berlin Gold Hat is covered in circular ornamentation representing lunar cycles. It was probably used to calculate intercalary months, also known as leap months, in order to bring the lunar year in alignment with the seasons of the Sun. The hat dates to the late Bronze Age, about 1000–800 B.C.E., or three thousand years ago.

This basalt stela depicts a king and the Star of Ishtar, or Venus; the winged disk of the Sun God, Shamash; and the crescent of the Moon God, Sin. The Moon God's representation is the closest to the king and the largest, depicting his supremacy. The stela's inscription recounts how King Nabonidus, the last king of Babylon, performed good deeds and as thanks, the gods put an end to a period of drought, bringing prosperity to his country. The stela was excavated near what is now Nasiriyah, Iraq, in the ruins of Babylon.

This calcite disk was excavated in 1926 from the temple of Ningal, revered as the wife of the Moon God Nanna, at the palace of Ur, an ancient city in what is now Iraq. On one side, seen here, is a depiction of a sacrifice being completed by priestesses including Enheduanna, daughter of Sargon of Akkad, the first ruler to unify the lands of Mesopotamia. The disk's reverse side contains an inscription of the priestess. Enheduanna used worship of the Moon God to help unify the lands in Sargon's kingdom.

The Great Ziggurat of Ur stands after four thousand years near what is now Nasiriyah, Iraq. Its construction was completed in the twenty-first century B.C.E. by King Shulgi and restored in the fifth century B.C.E. under King Nabonidus. Its exterior was restored again in the 1980s C.E. by Saddam Hussein. The site was part of an American air base during the Persian Gulf War and the later war in Iraq. <small>U.S. AIR FORCE PHOTO BY STAFF SGT. CHRISTOPHER MARASKY.</small>

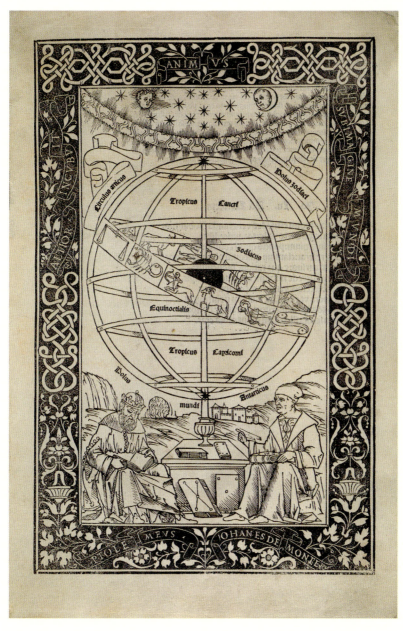

In this woodblock print from 1496, the man on the right, Johannes Müller von Königsberg, also called Regiomontanus, meets with Claudius Ptolemy. The centerpiece is an armillary sphere that depicts the night sky, including the constellations of the zodiac. The sphere represents Ptolemy's view of the cosmos, with Earth at the center of the universe. Regiomontanus co-authored an updated account of Ptolemy's masterwork, *Almagest,* which included simplified translations and commentary. The new version inspired a student named Nicolaus Copernicus to begin his own work, ultimately overthrowing Ptolemy's geocentric view and ushering in modern science.

Born Mikołaj Kopernik, the doctor Latinized his name after attending medical school. Copernicus revered Ptolemy as "that most astounding of astronomers," but was skeptical of the elaborate orbits Ptolemy's cosmology required. Copernicus's new cosmology, published in 1543 as *On the Revolutions of the Heavenly Spheres,* updated Ptolemy's ideas and would ultimately overthrow a millennia and a half of wrongheaded notions about the universe.

COURTESY OF THE SMITHSONIAN LIBRARIES.

This image was sketched on the night of July 26, 1609, by an English mathematician named Thomas Harriot. He was the first person to draw an image of the Moon as seen through a telescope, completing his sketch a few months before Galileo Galilei. The Moon was a five-day-old waxing crescent, which Harriot scribbled in a note at the top right.

PHOTO BY THE AUTHOR. IMAGE USED WITH PERMISSION OF LORD EGREMONT.

ctum daturam. Depreffiores infuper in Luna cernun-
tur magnæ maculæ, quàm clariores plagæ; in illa enim
tam crefcente, quam decrefcente femper in lucis tene-
brarumque confinio, prominente hincindè circa ipfas
magnas maculas contermini partis lucidioris; veluti in
defcribendis figuris obferuauimus; neque depreffiores
tantummodo funt dictarum macularum termini, fed
æquabiliores, nec rugis, aut afperitatibus interrupti.
Lucidior verò pars maximè propè maculas eminet; a-
deò vt, & ante quadraturam primam, & in ipfa fermè
fecunda circa maculam quandam, fuperiorem, borea-
lem nempè Lunæ plagam occupantem valdè attollan-
tur tam fupra illam, quàm infra ingentes quæda emi-
nentiæ, veluti appofitæ præfeferunt delineationes.

Hæc

Galileo Galilei trained as an artist and was capable of producing lovely renderings
of what he saw in his telescope. In *Starry Messenger,* from which this page is taken,
Galileo commented in detail on the Moon's apparent physical features, describing
the maria and the lunar craters, noting that some large spots appear "more de-
pressed" than brighter areas. COURTESY OF THE SMITHSONIAN LIBRARIES.

Johannes Kepler was a polymath, astronomer, devoted son, bereaved father, mystic, devout Christian, dedicated servant, creative thinker, beautiful writer, and scientific giant.

Wernher von Braun stands in front of the Saturn V rocket on July 1, 1969, as it is readied for the historic Apollo 11 mission. Von Braun, who dreamed of space exploration since boyhood and learned to build rockets under the auspices of Adolf Hitler, led development of the Saturn V at NASA's Marshall Space Flight Center.   COURTESY OF NASA.

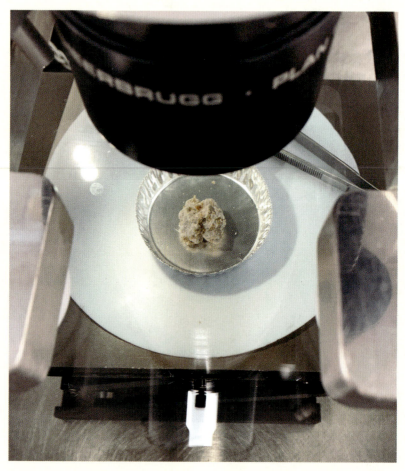

Troctolite 76535, a sample of the Moon collected during Apollo 17, rests underneath a microscope at NASA's Lunar Sample Laboratory inside Johnson Space Center. The rock has been dubbed the most interesting sample to come home from the Moon.

Astronaut Edwin E. "Buzz" Aldrin, Jr., climbs down the lunar module *Eagle,* preparing for his first steps on the Moon. Neil Armstrong, who stepped onto the Moon a few minutes before Aldrin, took this photo. For the first time, the human mind was occupying the Moon and looking upward and outward, rather than looking toward the Moon. COURTESY OF NASA.

Chimney Rock National Monument in southwestern Colorado is a natural alignment formed by erosion over eons. Every eighteen years, the full Moon rises between the spires. Around 1093 C.E., the site hosted a large city populated by a people often known as the Anasazi, now commonly called Ancestral Puebloans. They likely visited the site to commemorate the eighteen-year cycle known as the lunar standstill.   PHOTO BY THE AUTHOR.

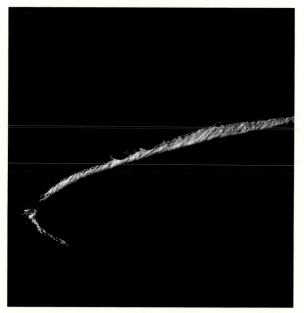

The Lunar Reconnaissance Orbiter captured this oblique view of the rim of Shackleton Crater, located at the Moon's south pole. Three points on the crater's rim remain sunlit for 90 percent of the year, while the crater interior never sees the Sun. Near the Peaks of Eternal Light are depressions on the crater rim that also sit in permanent shadow. These depressions form so-called cold traps that can contain ice, which private companies and government agencies are interested in harvesting.

NASA/GSFC/ARIZONA STATE UNIVERSITY.

Humans put away superstition* and the ancient world dissipated like a fog lifting, a fog obscuring the true Moon.

As it did under Plutarch a millennium and a half earlier, human understanding of the Moon took a giant leap forward. By the late seventeenth century, the Moon resolved into a three-dimensional place. Anaxagoras was right: It had mountains and valleys, and it was Earthy. Plutarch was right: This recognizable Earthiness could be mapped. The Moon had posed a problem in our cosmology for centuries, but when it finally became real to us, it revealed the truth.

○ •

THE RUPTURE BETWEEN the real world and the Holy Word finally cracked open a half century after the work of a quiet, pious university student christened Mikołaj Kopernik (he Latinized his name after attending college). But the origins of this schism go back much further, to the earliest stirrings of Christianity, to the Roman Empire, to the world's most magnificent library. A pioneering scholar sat under the eaves of the Library of Alexandria and paged through its scrolls, thousands of them. The library in the city of Alexander, the great conqueror who died in Babylon, had stood for centuries by the time the scholar worked there. Its collections weathered the sieges of Julius Caesar and his friend Mark Antony and Caesar's nephew Octavian, and the library was restored after Rome took over the Egyptian city when Octavian defeated Cleopatra and Antony in 30 B.C.E.

The scholar's surname, Ptolemy, echoed the Egyptian dynasty that ruled the region for centuries, but his first name was Claudius, a Roman appellation indicating his citizenship in the great new empire. Claudius Ptolemy was the only great astronomer of Roman Alexandria, and he used the library's contents to come up with a mathematically justifiable world order.

Before Ptolemy, a simple, Aristotelian, Earth-centered universe was the most widely accepted cosmology. The other planets, the Moon, and the Sun traced a circular path—because circles are a "perfect" shape— around Earth. The celestial bodies do not actually do this, of course, and

* To a certain extent, anyway.

that is obvious if you watch them. So philosophers performed mental gymnastics to satisfactorily "save the phenomenon" of the Sun's apparent rotation around Earth. A theory of cosmology must account for the actual cosmos, after all.

Claudius Ptolemy performed incredible feats of mental goaltending to achieve this save. And it is his work that made geocentrism so persistent in Western culture.

We know very little about his life, but he was probably born around 100 C.E. and lived in Alexandria. He was, from a young age, enthralled by astronomy. Working in that great library, with access to every important text ever inscribed up to that point, Ptolemy made himself one of the world's first true global scholars, marrying the collected knowledge of Babylon, Greece, Egypt, and Rome. Ptolemy's masterwork synthesizes a millennium's worth of Aristotelian physics, Archimedean math, Babylonian astronomy, Euclidean geometry, and his own contributions. It is an impressive feat of scholarship and original thinking, and it is utterly wrong.

Ptolemy called it *Mathematike syntaxis* (*Mathematical Arrangement*). But through the Arabs who kept it safe and translated it, the treatise became known by the first word in its Arabic title: The Greatest Compilation, المجسطي, or *Al-majisṭī. Almagest.*

Ptolemy's model of the heavens placed a stationary Earth at its center, just as thinkers before him had done through the ages. But Ptolemy was the first to attempt to explain how this arrangement could work, using what might generously be considered fuzzy math. He made a series of omissions and intentional cheats that attempted to explain the different movements of the Sun, Moon, planets, and stars. The most important is Ptolemy's description of an idea that dates to a few centuries earlier, from Hipparchus, the one who figured out the precession of the equinoxes: the epicycles. It's a convoluted way to work out circular orbits for all the planets, and it calls for a series of circles moving around the circumference of larger circles. It's like a Spirograph for the solar system.

In Ptolemy's system, the cosmos is arranged in a circle with an imaginary point at its center. To one side of that imaginary point, barely offset from the middle, sits Earth. On the other side of the imaginary point, opposite Earth, is another pretend location Ptolemy called the "equant." All the planets orbit Earth in a circular path called a "deferent." Each body that goes around Earth sweeps out from the equant, at equal angles, to move along its deferent. The equant is essentially a tool that enables the planets to move at different speeds along the deferent, because the planets do appear to move across the sky at different speeds. These are combined with a second orbital scheme, called an "epicycle." to explain retrograde motion. A planet's epicycle sits on the outside of its deferent. Imagine coffee-drip rings overlapping on a café table, forming concentric circles. Ptolemy posed that the epicycle spins counter to the deferent, so sometimes an object going around Earth would appear to slow down or move backward, thus explaining retrograde motion.

So now you have Earth inside a large circle that represents the movements of the other celestial bodies. They are also running laps along their own mini circles, which sit on the edge of the larger circle. Ptolemy's system of circles was thereby wrangled into a neat arrangement that predicted planets' locations with great accuracy. But the Moon was a problem for him.

To describe the motions of the Moon using these schemes, Ptolemy needed to calculate its distance from Earth and figure out where the Moon would orbit on its deferent. He used lunar eclipses to do this, because lunar eclipses were the best way for astronomers to make precise measurements of celestial positions. When the Moon is awash in Earth's shadow during a lunar eclipse, the Sun is, by definition, 180 degrees opposite the Moon's position. This is why the Moon appears red, as it reflects the sunsets of the day before and the sunrises of the day to come. When you know precisely where the Moon and Sun are in the dome of the sky, you can measure the locations of stars and planets, too. It's like checking a celestial compass to orient yourself.

Ptolemy consulted the Babylonian eclipse prediction catalogs and found a few upcoming eclipses that would have Sun-Moon arrangements of similar duration and geometry. They occurred on April 5, 125 C.E.; May 6, 133; October 20, 134; and March 6, 136. He also consulted old Babylonian eclipse records, which included the precise time of lunar eclipses.*

Ptolemy used the eclipses to come up with what amounts to a mathematical fudge: The Moon's deferent is tilted with respect to the others. But if this were true, the Moon's distance from Earth would change dramatically—by a factor of two or more—over the course of its orbit. At apogee, under Ptolemy's model, it would be twice as far away as at perigee. The marines at Tarawa Atoll would have seen a tiny half-Moon waning away, and been shocked two weeks later when it grew immense. This funhouse mirror Moon is not a real thing, of course. Though it does draw closer at times and present us with "supermoons," its apparent size is very similar, especially on the scales the ancients would have measured; on the Moon and its extremely complicated motions, the careful Ptolemy was not even close to correct.[2]

Islamic Arabs, who carried the light of consciousness through the Dark Ages, were apparently the only scholars to doubt Claudius Ptolemy's bizarre plan. From the ninth through early fourteenth centuries, many Islamic scholars recalculated some of Ptolemy's more cumbersome notions, improving on his planetary models. Some were completely unimpressed with him.

"Ptolemy assumed an arrangement that cannot exist," scoffed the Arab mathematician and astronomer Ibn al-Haytham (Latinized to Alhazen), the father of optics, in a withering critique called "Doubts Concerning Ptolemy." "The fact that this arrangement produces in his imagination the motions that belong to the planets does not free him from the error he committed in his assumed arrangement, for the existing motions of the planets cannot be the result of an arrangement that is impossible to exist."[3]

Ptolemy's crime, according to al-Haytham, is not just that his ideas were wrong, but that he ignored the problems inherent in his ideas, thereby denying that there could be any better ones. "In short, Ptolemy's approach is not the right path to be followed if science is to make any

---

* Ptolemy used all these Sun-Moon-Earth alignments to figure out the Moon's parallax. This just means a displacement in its apparent position in the sky.

progress at all," writes Hassan Tahiri, a twenty-first-century C.E. scholar of Arab science and philosophy in the Middle Ages.[4]

Outside the Islamic world, however, nobody seems to have asked many questions. *Almagest* and its complicated series of circles became the accepted explanation for the heavens above, and its ideas were incorporated into Western thought during the next fourteen centuries. With the help of a new, unquestioning Christian philosophy, Ptolemy's Earth-centered cosmos calcified, and for the most part, scholars left it at that.

We can thank Saint Augustine of Hippo, in part, for the hold Ptolemy's idea had over the Western conception of the world order. In the early fifth century C.E., Saint Augustine (354–430) looked inside Plato's cave and developed a new Christian philosophy. Plato had preferred the playground of thought over actual experiment because he believed our senses and our experiences deceive us of the truth—they are merely the shadows of reality. Augustine's Neoplatonic Christian philosophy likewise held that facts are merely illustrative of truth, not starting points for experiments that would lead to new questions. Augustine argued that ideal forms exist only in the mind of God, and that the reality that we can detect is just an illusion. To Augustine, the path to knowledge and enlightenment, truth and God, was reachable only by turning inward. Augustine's philosophy formed the pillars of the Christian church, and few would try to modify it* until the Enlightenment. For centuries afterward, the Church taught that truth was revealed only through Scripture.

Almagest fit nicely into this Neoplatonic worldview. At the very least, the prevalence of this school of thought helps explain why, for centuries, people accepted *Almagest* and hardly anyone thought to ask, "But what about the Moon?"

---

* The Franciscan friar Roger Bacon (namesake of the later Francis Bacon, progenitor of the scientific method) is one person who tried something different in the Middle Ages, attempting to discern truth from nature instead. For this brave effort, the elder Bacon was imprisoned in 1277 and his books were banned.

SCHOLARS TRANSLATED *ALMAGEST* from Greek into Arabic around 827 C.E., and after Christians conquered the city of Toledo in 1085, they translated it into Latin. *Almagest* remained the most popular astronomical text into the fifteenth century. But with the fall of the Byzantine Empire, a new intellectual era dawned.

In April 1453 C.E., the Turks laid siege to Constantinople, the seat of the Byzantine Empire for eleven centuries. The night of May 22, 1453, the Moon rose darkened and blood red, eclipsed. Seven days later the city fell to the invading Turks. As the empire fell apart, Greek-speaking scholars fled West toward mainland Europe, bearing with them their treasured texts, their translations of the works from antiquity, and their tremendous body of knowledge. One of these migrants was a man named Basilios Bessarion, who studied philosophy in Constantinople. He vowed to reunite the Greek and Latin worlds, both their religions and their faith traditions, and in this quest he saved huge swaths of material that may otherwise have been destroyed by the new Turkish regime.[5] Pope Eugene IV made him a cardinal, and Bessarion traveled the Latinate world, accumulating a huge library and befriending scholars. Among the texts in his collection was a copy of *Almagest*. Then, in 1460, Bessarion traveled to Vienna and had a fateful encounter with a young man named Regiomontanus.

Bessarion commissioned Regiomontanus and another scholar to produce a new translation of *Almagest*. It would be used for teaching, so they cut it in half and included a handy reader's guide—essentially a SparkNotes version of the most important mathematical treatise to come down from antiquity.

The *Epitome of the Almagest* was a pleasant abridgment of Ptolemy's masterwork, and like modern literary criticism, the translators had no problem pointing out errors, either of fact or of omission. Chief among Regiomontanus's critiques was the fact that Ptolemy had been wrong about the Moon. In 1491, an eighteen-year-old student at the University of Kraków picked up a copy of the *Epitome* and grew intrigued.

THOUGH COPERNICUS REVERED Ptolemy as "that most astounding of astronomers," he was skeptical of the elaborate orbits Ptolemy's cosmology required. They were messy. They were too complicated. They didn't make sense as a unified system of heavenly order, because each planet required its own set of finely tuned movements.

"I often considered," he wrote in 1514, "whether there could perhaps be found a more reasonable arrangement of circles."[6]

To figure it out, Copernicus turned to the Moon.

Like Ptolemy, Copernicus watched three lunar eclipses, in 1511, 1522, and 1523, using them as a celestial compass.[7] Then he used his eclipse observations to correct Ptolemy's lunar motions. Under the Copernican system, the Moon was not supposed to grow huge in the sky and shrink again over the course of the month. He knew that the Moon occasionally blocks the Sun, demonstrating that the Moon revolves around Earth. "In expounding on the Moon's motion," he wrote, apparently without irony, "I do not disagree with the ancients' belief that it takes place around the Earth."[8]

On March 12, 1529, Copernicus witnessed the Moon block the planet Venus from view: "I saw Venus beginning to be occulted by the Moon's dark side midway between both horns at one hour after sunset," he wrote in *On the Revolutions*. He used this occultation, and others from the star catalogs of antiquity, to deduce the motions of Venus. He was beginning to realize that Ptolemy was wrong about more than just the Moon's weirdly tilted deferent. He began to understand the truth: that the immobile Earth is not at the center of a whirling universe, but that we inhabit but one of many planets, which all revolve around the Sun.

Historians have long tried to figure out why Copernicus was struck with this revelation, which was completely unique as far as he knew.*

Some modern scholars suggest that Copernicus knew of research from thirteenth-century Islamic astronomers who were members of the so-called Maragha School. These astronomers worked at Maragha Observatory in what is now northern Iran, under the leadership of the Persian polymath Nasir al-Din al-Tusi.

---

* He wasn't aware of Aristarchus, who had promulgated a similar theory more than a millennium and a half earlier, because the writings of that Greek philosopher had not yet been reintroduced.

Al-Tusi and others came up with planetary models that eased some of Ptolemy's especially strained notions, including detailed geometrical solutions for the orbit of the Moon. Some of the math behind these models also appears in Copernicus's treatise, suggesting Copernicus might have known of the Maragha School's work. Although scholars still debate this history, it seems reasonable to assume that if Bessarion and Regiomontanus introduced *Almagest* to Europe, then other people brought different geometrical and mathematical works from the Arab-speaking world, too.[9]

We will probably never know; the chain of custody of such ideas may be all but impossible to recover. For his part, Copernicus writes that he was just bothered by Ptolemy's inelegance and, moreover, the system did not seem sufficiently divine.

"After long reflection, I began to be annoyed that the movements of the world machine, created for our sake by the best and most systematic Artisan of all, were not understood with greater certainty by the philosophers," he wrote in his book's dedication, which was to Pope Paul III. He mentions that he read the works of earlier philosophers, notably Plutarch, and noticed a few short references to other ancients who thought Earth moved around the Sun. He figured he might as well imagine what could be, he wrote to the pope.

Copernicus's groundbreaking theory had taken shape by 1514, but he did not publish it for almost thirty years, until he had reached old age. In the meantime, he worked as a priest and a doctor, but never neglected astronomy. He may have hesitated to publish because he understood the consequences of doing so. Even with the most religious, worshipful intent—even with a dedication to the pope himself—authoring a missive calling for Earth's displacement was revolutionary, even heretical. The very act of wondering about it was revolutionary. In Copernicus's world, repeated observation as a means of building knowledge, or what we would consider the definition of scientific pursuit, was rarely combined with the impertinence of asking hard questions.

"The men who founded modern science had two merits which are not necessarily found together," the historian of philosophy Bertrand Russell said of Copernicus. "Immense patience in observation, and great boldness in framing hypotheses."[10]

Anaxagoras had offered up some bold hypotheses, to be sure. And no one can doubt the incredible patience and reverent repetition of the sky priests of Babylon. But hardly anyone had both merits at the same time. "And," Russell says flatly, "no one in the Middle Ages possessed either."

But there was something Copernicus, like Anaxagoras and his fellow pluralists, still lacked: proof. There was no way to demonstrate Copernicus's bold claim, and no way to show that the heavenly bodies revolved around the Sun and not Earth. It would fall to the next generation to show their work and convince the world. And they would use the Moon to do it.

○ ·

THE OLDEST IMAGES of the Moon seen through a telescope are, frankly, bland. They are scrawled in ink on yellowing parchment and surrounded by notes, commentary on the weather, and other random scribblings. Though some are detailed, many are so vague you'd be forgiven for wondering whether they even depict the Moon at all. They were made by Thomas Harriot, someone whose sharp intellect and mathematical prowess did not necessarily translate to artistic skill, which is one of the reasons his drawings are not better known and he does not get the credit he deserves for bringing the Moon closer to home.

Crescent Moons, gibbous Moons, full Moons, sketched by Harriot, are found in a sheaf of paperwork stored unceremoniously in a storeroom. The storeroom is tucked in the basement of a grand country estate called Petworth House, one of England's proudest patrician mansions, located in a tiny village in West Sussex, fifty miles southwest of London. Upstairs, the palatial house is adorned with Romantic paintings by J. M. W. Turner and some of the most ornate furnishings imaginable. Downstairs and through a nondescript hall you can find a huge leather volume the size of an average cubicle desk, stuffed with Moons. Only Alison McCann can show them to you, and you can view them only with permission from one of England's aristocratic lords, the 7th Baron Leconfield, 2nd Baron Egremont, descendant of the Earl of Northumberland, known to history readers as the World War I scholar Max Egremont.

On a clear spring morning, McCann, Lord Egremont's family archivist, sat across from me in her office and poured us each a cup of coffee. The huge binder full of Moons would wait a moment; proper English etiquette insisted that we chat first. As I sipped, she told me about Harriot, English mathematician and scientist, and the creator of this bundle of Moons.

Harriot ought to be better known, because he was one of Elizabethan England's finest scholars. He is considered the father of modern algebra and at his time was regarded as the equal of any Renaissance thinker. He might be the greatest mathematician Oxford has ever produced. He served as a navigator and cartographer to Sir Walter Raleigh, one of Queen Elizabeth I's eminent transatlantic explorers. Harriot was one of the first Englishmen to make the journey to the New World, sailing to the doomed Roanoke Colony in modern-day North Carolina. His *A Briefe and True Report of the New Found Land of Virginia* might be the earliest European account of Native Americans, and of the continent's flora and fauna. He learned and translated the Carolina Algonquian language and learned from the Algonquin people how to smoke dried tobacco leaves. Harriot's writings helped make tobacco trendy among the elite, and its tangy, yellowy smoke soon spread beyond taverns and inns.

AROUND 1590, HARRIOT and Raleigh befriended a younger man, a mildly deaf, dark-haired, aquiline-nosed, would-be alchemist known as Henry Percy. He was the 9th Earl of Northumberland,* the heir to a prominent Catholic family who had run into trouble with the monarchy. Percy's father, the eighth earl, had committed suicide in the Tower of London, where he was imprisoned after being accused of conspiring to replace Queen Elizabeth with Mary, Queen of Scots. By 1598, Harriot was working for the charismatic Percy, known as "the Wizard Earl," who paid him a lifetime pension and gave him rooms in Syon House, at the Northumberland estate on the Thames. Harriot was able to live a scholar's life, and he didn't even have to publish.

"He had a house, he had a salary, and he could just indulge in pure science," McCann told me.

---

* Northumberland's outcrops of complex rocks formed during the closure of the Rheic Ocean, when fish learned to walk. Everything comes back to geology.

Harriot's breakthroughs mounted quickly. In 1601, Harriot discovered the law of refraction, which predicts how much light will bend when it travels through a material. The law had eluded Harriot's contemporary and frequent correspondent, the inimitable German astronomer Johannes Kepler. "Measuring refractions, here I get stuck. Good God, what a hidden ratio!" Kepler wrote ruefully in 1603.[11] Figuring out the law of refraction should have won Harriot enduring fame, but he didn't seem to care much about that. His former student and friend Sir William Lower berated him for his "too great reservednesse," arguing in a 1610 letter that Harriot should take credit for his discoveries, particularly his work in algebra. Harriot's decorum "hath robd [him] of these glories," Lower wrote.*

Harriot was a painstakingly thorough scientist, long before that word was invented, and made measurements in careful, repeated experiments that ultimately led him to new conclusions. He observed sunspots and noticed they appeared to be moving west, and used their appearance to correctly deduce the Sun's rotation period, though at the time nobody thought the Sun rotated at all. Though Sir Francis Bacon gets credit for originating the modern Western method of experiment-based science, Bacon himself said in 1608 that he was inspired by Harriot, who was "already inclined to experiments."[12]

In early 1609, Harriot got his hands on a new invention, which the Dutchman Hans Lippershey had tried to patent in October 1608. Spectacle makers were grinding glass in larger sizes and combining them to produce the world's first telescopes. Harriot bought one that magnified his sight by six times, and had it mounted on the roof of his quarters at Syon House.

On July 26, 1609, he became the first person to draw an image of the Moon as seen through a telescope. The Moon was a five-day-old waxing crescent, and Harriot placed it at the top of his page, the crescent arcing over the parchment.

I stared as, with gloved hands, McCann removed this drawing from its red folder. The first telescopic Moon image in human history looked

---

* Credit for the law of refraction went to Willebrord Snellius in 1621 and René Descartes in 1637.

very much like it does on the Nebra sky disk: thicker than a scythe, bulging, growing larger but not yet half full. It is not a beautiful image. Awkward and jagged, it seems as if its creator were unsure of his hand. Harriot drew a full circle, but sketched a ragged arc, forming the bridge of the nose of the Man in the Moon, to mark the boundary between the illuminated crescent and the darker part of the sphere. The full circle may have been visible through reflected earthshine, but it's not clear if Harriot wanted to convey that. He sketched some furry-looking features in its midsection, but they are vague and poorly rendered, not really representative of the familiar Moon.

McCann set the drawing on a cloth, and I craned my neck, trying to view it from a new angle. I thought about Vince Gaffney suddenly seeing the calendar of Warren Field, and I pulled up a modern Moon map on my phone. Comparing the two images, I could see that Harriot's furry features might depict the Sea of Tranquility, the spot where Apollo 11 would land three and a half centuries later.

This image predated the drawings of that talented Tuscan, Galileo Galilei, by four months.

To celebrate the four hundredth anniversary of the first telescope use, in 2009, the Egremont estate shared the image with the British press, and McCann was inundated with queries. She was frustrated by the attention and the sudden realization that Harriot, not Galileo, was the first to view the Moon through spyglass. The Egremont family had known the whole time. "It really annoys the Italians. They borrowed this for an exhibit about Galileo, and I bought a copy of the catalog because it's such fun how dismissive they are of Harriot," she added with a grin.

The drawings are monumental not because they are good, and not just because they are first, but because of what they are. Harriot was the first person to record the Moon—or any astronomical object—through a telescope. He was also the first person to grapple with what that meant. The telescope had finally let humans slip the surly bonds of Earth[13] and see beyond it, beyond ourselves, even. What Harriot saw defied the imagination, even philosophy. For Harriot, there were no introspective ramblings on the Moon's features, no fervent or religious embellishment of its chalky face. He did not even emulate Copernicus, his more recent forebear, and attempt to ascribe any scriptural beauty or heavenly order. Harriot was educated in the classics and he knew Plutarch's *De Facie,* but even his distant Moon-watching kin could not prepare him for actually *seeing* literal terrestrial features on another world. It must have astounded him. At no point in human history had anyone ever seen the Moon magnified the way Harriot now could. To document it, perhaps to share it with others, Harriot rushed to record what he saw in the form of a picture. He must have known, even though he would never publish them, that his uncertain scribbles would be more valuable than any words he could jot down.

Galileo knew this, too, and, trained as a painter, he did a much better job conveying what he saw. Galileo's masterwork *Sidereus nuncius* (*Starry Messenger*) was published in March 1610. Harriot received one of the first copies, hot off the press,* and read it avidly. Correspondence between Harriot and his friend Lower show their mutual excitement about Galileo's work and their own enjoyment of the telescope, which they call "the perspective cylinder." At one point, Lower asks Harriot to "send me also one of Galileo's books if any yet be come over and you can get them."[14] As a scholar of renown, in frequent contact with men like Kepler, it's likely he received books like *Starry Messenger* and Kepler's *Conversation with the Starry Messenger* from the authors themselves.

McCann showed me how Harriot's own drawings progressed after he saw what his Italian counterpart had produced. On some pages, he drew crescent Moons side by side, or small half-Moons accompanied by algebraic tables. He apparently liked to plan ahead, because McCann showed me several pages that had been marked with a perfect circle, ready to be used as a Moon template, though Harriot didn't use them all. One page in particular may have made the distant ancestors of the Percy family, the Scots of Warren Field, quite happy. It shows a sequence of twelve circles, each with a different Moon phase, from crescent to gibbous to full.

On September 11, 1610, Harriot sketched a thin crescent with a very prominent dimple, which again may have been the Sea of Tranquility.

---

* The printing press, invented around 1440 C.E., had become commonplace by the early seventeenth century.

"The appearance was notable, rugged in many places," he wrote, complete with "ilands and promontoryes." "But I could not get down the figure of all, because I was troubled with the reume," he concluded. He had a cold and went to bed.[15]

Harriot also produced the first drawing of a completely full Moon. Finished in 1611, it is heavily annotated with letters and numbers, which Harriot used to mark his position as his eyes darted from the eyepiece to the page. It is divided into light and dark, maria and plains, an unpretentious-looking and mathematically precise diagram. He was a mapmaker, after all, and his drawings were a form of cartography. He even wrote down his telescope's magnification, like a sort of map key. By 1613, Harriot had produced two full-Moon maps, many with features you could easily recognize in today's high-definition imagery.

Though the drawings are drab in comparison to Galileo's, they are nonetheless extraordinary. They represent a forward shift in thinking that married raw observations with accurate mathematical interpretation.[16] Harriot's maps would have amazed Plutarch and Ptolemy, not to mention the Neolithic Moon watchers of Scotland. His maps were especially impressive given the rudimentary telescopes he worked with. The level of detail in his full-Moon map would be unequaled for another half century, until people started grinding high-quality telescope glass, leading to a seminal lunar map by the Polish cartographer Johannes Hevelius in 1647.

WHEN IT CAME time to document the first-ever close-up of the Moon, pictures were all Harriot managed. It wasn't as though Harriot couldn't write; in the mysterious Roanoke Colony, he recorded the flora, fauna, and native North American inhabitants in assiduous detail. But he did not attempt to offer a high-level interpretation of his observations, nor were his Moon drawings accompanied by any attempt to rearrange the heavens. Often, he didn't even annotate them. Because he didn't publish, Harriot's diary entries are the only version of his history with the Moon. We can't know what Harriot wondered as he stared at the large bright crescent through his spindly brass spyglass. We don't know why he never published his drawings. Biographers have suggested that

Harriot cared more about living his life than about attaining fame. He may have hesitated to invite more scrutiny upon himself because he was an atheist, or at least suspected to be one. He never had a financial need to publish; Raleigh and then Percy kept him gainfully employed, and he had no reason to impress any benefactors or religious elite, unlike his contemporaries Kepler and Galileo. But it may have been political drama that kept Harriot quiet, and that may explain why his papers wound up at Petworth House.

Around 1597, when Harriot first went to work for Henry Percy, Queen Elizabeth I refused to name the Scottish king James VI as her heir. This set off ferocious infighting among factions who fought to succeed "the Virgin Queen" as the ruler of England and the empire. When she died at age sixty-nine, in 1603, James became king anyway. The palace intrigue made the new monarch view many elites, including Raleigh and his friends, with suspicion.

Then, on November 5, 1605, conspirators including Guy Fawkes, Henry Percy's relation Thomas Percy, and several others were arrested for attempting to blow up the Houses of Parliament. The infamous Gunpowder Plot led to a series of side investigations, and even Harriot himself was briefly imprisoned, accused of casting a horoscope of King James as a form of dark magic. Eventually, Henry Percy, Harriot's friend and patron, was imprisoned in the Tower of London. He remained there for sixteen years, as Harriot studied optics, saw Halley's Comet, and mapped out his observations of the Moon.

After Percy was released, he was barred from living in London and retired to Petworth House, McCann told me. There, he received Harriot's papers after the eminent scientist died of cancer on July 2, 1621.* Egremont, Henry Percy's distant descendant, is now the steward of 453 pages of Harriot's work. McCann knows them all well.

"I don't think I've ever got blasé about it," she told me. "What I love is actually the feeling that you can almost make these people live on. They

---

* Scholars believe Harriot probably suffered from lung cancer. He may have been the first Englishman to die of lung cancer after a lifetime of smoking tobacco, which he helped introduce to mainland Europe.

are only little glimpses; if you think logically about what's left, it's not a complete record. It's not every bit of paper that was ever created. But you get these little glimpses of people's lives."

I thought about Harriot on my train back to London. The Sussex countryside and its lemon-yellow rapeseed fields melted into steel and glass as my train rumbled toward the great city. Inside Petworth House, surrounded by Turner's paintings and the grandest, most absurdly ornate furnishings I have ever seen, I might as well have been a courtier in the seventeenth century. In London two hours later, I breezed through tube stations, climbed onto Kensington High Street, and stood at a counter to drink an espresso. It was jarring, like traveling into the future.

Did Harriot feel something similar, squinting through his 6x magnification lens? Harriot and Galileo, Copernicus and Kepler, all these brothers in science, must have felt the weight of their discoveries; they certainly knew they were charting a new course. They talked about it with one another in letters, offering praise in their correspondence, and clearly enjoyed the adventure of discovery. But did they know they stood on a precipice between two worlds? Did Copernicus really understand that his certainty about the "chief world systems," as Galileo called the heliocentric and geocentric models, would upend society as he knew it? For that matter, did Galileo?

Even if they did not fear the truth, the reality of the Moon must have been unsettling. Through observations that began with Harriot, it became clear that the Moon was not made of "ether" or some other substance. It was rocky, mountainous, cratered, and contained dark areas that looked like seas. It was a *world*. It was, like brave Anaxagoras had said, "Earthy." Harriot saw it this way himself. McCann showed me where Harriot named one feature "The Caspian," which appears isolated on his drawing, as the most "land-locked" watery feature, just like Earth's Caspian Sea. The Caspian corresponds to the area later named Mare Crisium, on the near side of the Moon's northeast corner, northeast of the Sea of Tranquility. Mare Crisium means "Sea of Crises."

It's an apt name for the only lunar feature Harriot cared to label. He observed the Moon at a time of crisis in his own country and in the broader Western world. While Harriot stared through his perspective cylinder, Galileo's observations and astute writing upended the Aristotelian, ethereal order. Galileo's beautiful drawings went a step beyond Harriot's. He illustrated what he saw, making a valiant effort to show others how things really were, but on top of that he offered his own philosophical interpretation.

○ ˙

A HALF CENTURY after he published them, Copernicus's ideas would change everything for both Galileo and a young Johannes Kepler, at university in Germany.

In 1595, Kepler dove into Copernicus's work in much the same way that Copernicus had studied the *Epitome of the Almagest*. On a scholarship to learn philosophy at the University of Tübingen, Kepler read Plutarch's *De Facie* and Copernicus's *On the Revolutions,* and proclaimed that the Copernican system was "an inexhaustible treasure of truly divine insight into the wonderful order of the world and all bodies therein."[17] Kepler decided to focus his dissertation on the Moon, aiming to use its motions to prove Copernican ideas. Amid his studies, he emulated his ancient predecessors and used shadows to determine the height of some lunar surface features, as he would later explain to Galileo.[18]

Around the same time, about 475 miles to the south, at the University of Pisa, Galileo was employed as a mathematician. In his study of gravity and motion, he gradually came to doubt the teachings of Aristotle. His uncertainty about Aristotle led him to question everything else, and, like Kepler, to accept the wild heretical ideas of Copernicus. In the summer of 1609, while Harriot looked through his telescope and made his sketches in England, Galileo visited Venice and, for the first time, gained access to a telescope. There, he observed the Moon's surface; the cloudy, luminous stripe of the Milky Way; and the large moons of Jupiter, which he called "the four planets."*

In *Starry Messenger,* Galileo draws a distinction between the maria and the lunar craters, the latter of which "have never been observed by any one before me."[19] He correctly explains, as Harriot had roughly sketched, that the night advances on a jagged edge across the face of the

---

* Jupiter actually has at least eighty moons, but four was still a big surprise.

Moon. The shadows of the Moon were present because the Moon had surface features. This was a stunning departure from Aristotle's view, long believed, that the Moon was a spotless sphere, though it echoed Plutarch's assertion that the Moon was Earthy. Galileo still stops short of calling the surface features mountains, referencing vague "prominences and depressions," and is careful to avoid saying the Moon is like Earth. But if it was, he thought the dark spots were seas: "if one wishes to revive the ancient opinion of the Pythagoreans, that the Moon is like another Earth, the brighter part of which represents the surface of the land, and the darker part more fittingly the expanse of water."[20]

Though Galileo's telescopes were modest by modern standards, thin brass instruments that magnified his vision some twentyfold, they were more powerful than Harriot's spyglass. Now in a museum in Florence, mounted in glass cabinets with carefully controlled humidity and temperature sensors, they are some of the most precious pieces of technology in the history of humanity. Even Galileo's contemporaries understood the weight of his ideas.

"Most accomplished Galileo, you deserve my praise for your tireless energy," Kepler gushed to him. "Putting aside all misgivings, you turned directly to visual experimentation. And indeed by your discoveries you caused the sun of truth to rise, you routed all the ghosts of perplexity together with their mother, the night, and by your achievement you showed what could be done."[21]

In the years after publishing *Starry Messenger,* Galileo wrote letters on the nature of sunspots and conducted experiments to understand gravity. In 1632, he finally decided to circulate his own argument for the moving Earth, *Dialogue Concerning the Two Chief World Systems*. In colloquial Italian, not Latin, he outlines the case for Copernicus while attacking the demonstrably wrong physics of Ptolemy and Aristotle. While Harriot was content to contemplate the Moon from his roof and publish nothing, Galileo got global attention because he sought it. He wanted a broader audience for his ideas.

In his dialogue, he argues that the Moon and Earth are both clearly round. He claims the Moon is solid and dark, like Earth—meaning not made of ether. He correctly speculates on the nature of earthshine, which people had for centuries named "the ashen glow," the phenomenon that lets you view the dusky figure of the dark sector of the Moon. This is visible when the Moon is a crescent. It happens because sunlight reflected from Earth's surface is bright enough to partially light the Moon, too, like a mirror shining back at you. In a scientific echo of Enheduanna's paean to the Moon 3,860 years earlier, "*Your shining horns are...holy and lustrous,*" Galileo wrote that "this light is seen most clearly when the horns are the thinnest."[22]

All these impressions marked a giant leap forward from the old Aristotelian view of observation for its own sake. It was also a giant leap from the Neoplatonic view of accepting what you were taught, especially Scripture, at face value, over the observations of your own eyes.

Galileo followed the ancient Greek tradition and wrote his argument as a conversation, in which his characters talk about the two systems and debate their usefulness. It's a storytelling framework that makes scientific arguments easier to digest, but it was also strategic. In dialogue form, Galileo himself didn't have to take sides, or discuss Copernican ideas on their merits. But it is obvious where he stood. His Aristotelian character was named Simplicio, for instance. Not exactly subtle.

After a discussion of the Moon's orientation in the sky, another character, Salviati, says he's glad Simplicio finally gets what he is trying to say.

"For my part I more often encounter heads so thick that when I have repeated a thousand times what you have just seen immediately for yourself, I never manage to get it through them," Salviati says.[23]

What came next will always remain one of the most famous trials in the history of the Western world. Galileo was charged with heresy in 1633 but avoided being burned to death at the stake because he agreed to plead guilty and deny heliocentrism. He spent the rest of his life under house arrest.

As far as the Moon was concerned, however, Galileo's impact was irreversible. His Moon illustrations, like Harriot's, fulfilled the words of Anaxagoras and the prophecy of Plutarch: The Moon was Earthy, and it was flawed. Galileo's artistic renderings look natural, real, and artistic. They are pure art, "a self-contained embodiment of observation," as the historian Scott Montgomery put it. The Moon was teleported to Earth, and the truth of its image was revealed through the lens of a telescope, not through the written Word.

One of Galileo's lifelong friends, Lodovico Cigoli, even used the newly realistic Moon to transmute religious art, in the very seat of power of the Church.

In 1612, two years after Galileo stunned the world with *Sidereus nuncius,* Cigoli completed a fresco in the Basilica of Saint Mary Major in Rome, the most important Marian shrine in Christendom. During Galileo's lifetime, the large basilica had two new chapels added* and Cigoli was hired to paint a scene from the Book of Revelation, which takes place during the Apocalypse. His subject was "a Woman clothed with the Sun, and the Moon under her feet," standing opposite Saint Michael the Archangel. Cigoli painted as he was told. But the Moon at her feet was not the smooth orb of many early counterparts, who used the spotless Aristotelian Moon as a metaphor for the spotless Virgin Mary, mother of Christ.

It was Galileo's Moon, craggy, crater-pocked, indented, marred. Cigoli painted an astronomical Moon, not the Platonic ideal of a Moon. This fresco is still the subject of debate among scholars, but whatever Cigoli's intent—acting contra Counter-Reformation Church politics, or just making a very well-placed hat tip to his art school friend—his Moon changed things.[24] Moons in art became real. Paintings might have remained religious in nature, but their depictions of the sky and Moon were vivid, believable, and terrestrial, taking their cue from modern astronomy.

THE MOONS OF Jupiter were another huge problem, both for the world order and for the Moon. By charting the Jovian satellites' motions, Galileo had shown that not all heavenly bodies revolve around Earth. Io, Europa, Ganymede, and Callisto—now called the Galilean moons—

---

* The newer one was the Pauline Chapel, built by Pope Paul V, and the older one is called the Sistine Chapel, built by Pope Sixtus V. Though it has the same name, this Sistine Chapel is not to be confused with the one in the Vatican, the one painted by Michelangelo a century earlier.

were proof of the Copernican principle and challenged Ptolemy, Aristotle, and the Church.

But the addition of these new worlds also demoted our own Moon. Our satellite was no longer perfect or unique, and therefore neither were we. Earth was not even the only world with a companion. The Moon was toppled from its place of honor, no longer on par with the Sun and the observable planets. It was demoted to humble satellite, and just one of many.

"Our moon exists for us on the Earth, not for the other globes. Those four little moons exist for Jupiter, not for us. Each planet in turn, together with its occupants, is served by its own satellites," Kepler wrote to Galileo,[25] using the word "satellite" for what may be the first time. The word is rooted in the Latin *satelles,* meaning "attendant."

Other worlds were also served by their satellites. They were worlds accompanied by worlds. This is obvious now, but it was an extraordinary insight considering the time in which these people lived.

There was no broader universe beyond the solar system. In the middle of the second millennium C.E., there were still no such things as galaxies. Nobody knew how the universe was arranged. When Copernicus was in medical school, he was educated in techniques designed to balance "the four humors" of blood, phlegm, black bile, and yellow bile, and learned astrology to aid in the diagnosis and treatment of disease. In medicine, the phases of the Moon in particular were thought to be correlated with severity of illness. Most people still believed, as they had for centuries, that the Moon was a smooth sphere and that Earth was the center of everything.

Galileo's pronouncements to the contrary—backed by evidence—were a shot heard around the world. They were a salvo in a battle that had been brewing since at least the days of Copernicus, since Regiomontanus translated *Almagest* with a skeptical eye. In this battle, received authority was pitted against scientific inquiry. It set the written Word of Scripture against the evidence that lay before our own eyes: "Tremble before him, all the earth; he has made the world firm, not to be moved" (1 Chronicles 16:30). Galileo's artwork and his writing flew in the face of the Council of Trent, the great Catholic reply to the Lutheran

Reformation, which proclaimed that "no one, relying on his own skill, shall . . . presume to interpret the said sacred Scripture contrary to . . . the unanimous consent of the Fathers," even if those interpretations were new and never intended to be published.[26]

Many historians consider this schism to mark the dawn of modern science, with the Moon in a starring role. The scientific revolution certainly gained momentum in the decades to come, but the break between religion and empirical science was not a clean one. The change took place on a continuum. As Galileo understood, the Moon was still powerful as a symbol, and as more than a mere chink in the scriptural armor. Galileo's painterly views could not divorce the Moon from the divine, nor remove it entirely from myth. Exploration could not kill spirituality. Galileo's frequent correspondent, Johannes Kepler, understood this better than arguably anyone else of his generation. Though he revolutionized astronomy by giving us the laws of planetary motion, Kepler imagined that the heavenly bodies produced music, harmonies, as they moved around, and he was unafraid to attribute meaning to the tunes. Though his conclusions are scientific, his writing is imbued with the mysticism of the medieval world he inhabited.

After spending millennia as a perfect sphere representative of the divine, the Moon became Earthy in both a literal and a figurative sense. Its role above us had changed dramatically in the two millennia after Anaxagoras, and once again in the century between the books of Copernicus and Galileo. We had unmasked the Moon as a thief of the Sun's light, and as just one rocky companion among many satellites of many planets. We mapped its surface in glorious, illuminated detail, and we named its features. It became a place where we could imagine ourselves, because that was fun and newly possible, but also because it was easier, maybe even safer, to play out our human struggles someplace else.

In the centuries to follow, through art, music, and literature— including the first work of science fiction, by Kepler himself—the Moon would emulate Galileo's ashen glow, reflecting us back to ourselves.

CHAPTER TEN

Journeys of the Mind

IT WAS A dark and stormy night.

It was 1608, and no one had yet seen the Moon through a telescope. But it was up and full, and after pondering it for some time before the clouds rolled in, Johannes Kepler fell into a deep sleep and began to dream.

In his dream, he picked up a book by an Icelandic astronomer named Duracotus, who explains how long ago, he had ticked off his mother by peering into a mysterious parcel she was planning to sell to some sailors. She kicked him out at age fourteen, and he sailed off to Denmark. There, he met the famous astronomer Tycho Brahe, who took young Duracotus under his wing and taught him all about astronomy.

After several years studying with the genial Dane, Duracotus returned to his homeland and was reunited with his mother, Fiolxhilde. And it turned out she knew about astronomy, too. But she didn't need Brahe's "wonderful machines" to know the face of the Moon. Instead, she used magic to converse with the Moon just as clearly as if she were walking on it. She knew how to summon nine spirits from the tribes of beings who lived up there, on an island they knew not as our Moon, but as Levania.

Fiolxhilde decides to teach her son her ancient ways, and the two summon a tutelary daemon from Levania.* He is "most gentle and pur-

---

* This is not an evil demon in the way modern Western people think of them, but more like a spirit guide.

est of all, [and] called forth by twenty-one characters," the mother explains.

In a hoarse, lisping voice, this daemon holds forth on several subjects, from the Metonic cycle—that nineteen-year round known to the ancients, corresponding to the time it takes for the Moon and Sun to line up again in the same spot in the sky—to the geology and biology of his home. According to the daemon, Levania is covered in mountains, oceans, and caverns, and the life-forms that dwell among them are all huge and short-lived. They call Earth Volva, because it clearly revolves around Levania. "The most pleasant of all occupations on Levania," Duracotus explains, "is the contemplation of its Volva."

The Levanians build boats and crawl through caves to escape the scorching heat of the lunar daytime—equivalent to fourteen Earth days, Duracotus recalls. And they hide underwater when the Sun makes it too bright. "By combining nature with art, they can take refuge," the daemon says.

At this, hearing lashing rain and thunder, Kepler woke up. The tale comes from his book *Somnium, seu astronomia lunari* (*Dream, or Astronomy of the Moon*), circulated during his life but only published after his death.

○ ˙

JOHANNES KEPLER STOOD on a fulcrum between two worlds. His training, education, upbringing, and belief systems were decidedly medieval, but his own research and writing were practically modern. He learned Ptolemaic astronomy, read the ancient tablets and star catalogs, but then he performed measurements of his own in order to draw independent conclusions. At the University of Tübingen, he became a believer in the Copernican system and even tried to argue for it publicly, but the university wouldn't let him. His beloved professor Michael Maestlin was ultimately forbidden from speaking openly about Copernicus. Though Kepler shelved his dissertation, which focused on the Moon's motions and was aimed at proving Copernican principles, many of its ideas would be expressed in the book that became *Somnium*.

Like his contemporaries, Kepler was Moon-obsessed, and this obsession led him to some of his greatest insights. He is remembered as the legislator of planetary orbits, but Kepler was also the first to suggest that the Moon is responsible for the tides.

Kepler posited in 1609, in his groundbreaking *New Astronomy*, that the Sun is the center of the orbits of the planets; that the Sun moves them along their orbits, and that Scripture, which states otherwise, should be appreciated as poetic analogy but not taken as dogma; and that the planets do not orbit the Sun in a circle, but in an ellipse. He wrote that he used to think the planets were alive and moved around because they had souls, but that physics makes more sense.

In this beautifully composed work, Kepler also argued, for what is likely the first time, that the Moon's gravity causes the tides. He considered magnetic attraction, and reasoned that attraction is caused by two bodies rather than one. "If the Earth ceased to attract the waters of the seas, the seas would rise and flow into the Moon," Kepler wrote. "If the attractive force of the Moon reaches down to the Earth, it follows that the attractive force of the Earth, all the more, extends to the Moon and even farther."[1]

Galileo found this absurd and mocked lunar tides as "childish" and "occult" in 1616's "Discourse on the Tides," but of course Kepler was right.*

Kepler's findings were all remarkably bold, especially for a devout, mystical man writing in the early seventeenth century. Kepler stood above the field of science arrayed before him, from Copernicus to Brahe to Galileo, and applied his own observations to draw original conclusions. Kepler's creative thinking and his novel ideas position him, like Anaxagoras, as a bridge between the old world and the new.

GALILEO WAS PERSECUTED for his work proving the Copernican perspective, but he gave people a chance to see for themselves the empirical truth of his findings. Along with posing a serious problem for the Renaissance-era Church, the observable fact of actual worlds and their own moons, plural, generated lots of speculation about their nature and

---

* Sir Isaac Newton finally figured out the full tidal mechanism eighty years later. In the *Principia*, he correctly showed that the tides depend on both the gravitational attraction of the Moon, like Kepler said, and centrifugal forces caused by the Moon-Earth orbit.

their purpose. Kepler himself mused in correspondence with Galileo that one day someone would build ships to "sail the void between the stars." He was sure there would be plenty of explorers ready to volunteer for the journey. "In the meantime, we shall prepare, for the brave sky travelers, maps of the celestial bodies—I shall do it for the Moon, you Galileo, for Jupiter," the imperial mathematician wrote.[2]

This was a brief hint at *Somnium,* which Kepler had already completed in secret but shelved. Kepler told Galileo that the manuscript was a type of "lunar geography." Isaac Asimov and Carl Sagan both considered it the first work of science fiction. The book is simultaneously a manifesto for science that went against Church orthodoxy and an ode to the mythical, beautiful, spectral Moon. In straining against the accepted geocentric view, which Kepler knew to be wrong, he spins a tale of literal flights from Earthly reality.

A SMATTERING OF earlier works also feature imaginary lands, but they don't go nearly as far as Kepler and they don't invent fantasy for the purpose of explaining reality. Plato wrote about a mysterious lost city called Atlantis in his *Critias,* a sequel to *Timaeus,* but the story is not really based in science. A couple centuries later, Plutarch's *De Facie* asked what types of inhabitants the Moon might harbor, but the story includes no make-believe visit. A few Greek-speaking poets, notably the Syrian poet Lucian, also wrote about people going to the Moon, but their works are intended as satire. As pieces of fiction, *Critias* and *De Facie* serve as a scaffolding for philosophical argument, but they're not really allegorical.

Somnium depicts a strange and unfamiliar world, but it grounds this space oddity in reality by relating it to scientific knowledge. It's the first time someone imagined the Moon as a place inhabited by strange beings that could talk about what it was actually like, based on human observations. While the book is scientific in its tone and purpose, it also reveals that the Moon held deep spiritual significance for Kepler himself. He was able to hold two opposing ideas at the same time: the Neoplatonic view, that the Moon is a pit stop for souls,* and a more modern

---

* In a nod to Aristotle, Plutarch, and Saint Augustine.

view, that the Moon is a body with mass, and that it obeys the laws of planetary motion—the latter of which Kepler himself discovered. This duality is a sharp break from classical tradition, and it represents a new, modern way of thinking. For the first time, storytelling is used to advance modern scientific speculation.

Kepler's Moon has two hemispheres, called Subvolva and Privolva. Subvolva is the near side, the half of the Moon that faces Earth always, while Privolva never sees Earth. This Moon is also covered in pine cones that release living creatures when the conditions are right. In *Somnium*, travel between Earth and the Moon takes just four hours, but is only possible during a lunar eclipse, and is exceedingly dangerous for humans. Kepler reasoned, long before anyone understood the mechanics of flight or even Newton's laws of motion, that overcoming Earth's gravity would take extraordinary force. Daemons must help anesthetize any Moon-bound traveling humans so their bodies wouldn't fall apart after being "spun upward as if by an explosion of gunpowder." This detail would blossom in the mind of Jules Verne, Kepler's successor in imagination based on lunar realism. It would ultimately guide a German rocket scientist named Wernher von Braun and his contemporaries on their quest to ignite real rockets.

Between this strange combination of mysticism and hard science lies the real goal of *Somnium:* to attack the prevailing, Ptolemaic worldview that Earth is the center of everything. *Somnium* scoffs at the idea that people can be assured this is true simply because we can see the Sun and stars circling us. Kepler argues that the same thing would be true anywhere else. The Levanians see Earth and the Sun moving across their sky, so they assume their home, the Moon, is the center of the universe. And why wouldn't they? Everyone believes they're the center of the universe. We all, at some point, experience the tragic epiphany that demonstrates otherwise. "Of all discoveries and opinions, none may have exerted a greater effect on the human spirit than the doctrine of Copernicus," another German, Johann Wolfgang von Goethe, would say two centuries later. "Never, perhaps, was a greater demand made on mankind."[3]

The narrative in *Somnium* is much shorter than Kepler's 223 footnotes and charmingly titled "Geographical or, if you Prefer, Selenographical Appendix." The footnotes, added between 1622 and 1630, are heavy with science, and modern scholars believe they were partially meant to exonerate Kepler's mother, Katharina, from accusations of witchcraft, an eventual outcome of Kepler's attempt to gently awaken Earthlings to our centuries-long sleep.

"Everyone says it is plain that the stars go around the earth while the Earth remains still," he wrote in footnote 146. "I say that it is plain to the eyes of the lunar people that our Earth, which is their Volva, goes around while their moon is still. If it be said that the lunatic perceptions of my moon-dwellers are deceived, I retort with equal justice that the terrestrial senses of the Earth-dwellers are devoid of reason."[4]

IT'S HARD TO imagine how Kepler could have completed *Somnium* without a telescope. Unlike Plutarch and Lucian, he could actually see the Moon in great detail, and he could tell, as Galileo had explained, that the Moon was a world. *Somnium,* like Kepler himself, is a bridge between the old ways of thinking and the new.

"It has these postmodern features in a medieval text, and it's bringing together all these motifs from different kinds of genres," Dean Swinford, an English professor at Fayetteville State University who is an expert on *Somnium,* told me. It is proto–science fiction, but it is framed as a dream, a familiar storytelling structure in both medieval and classical literature. A dream narrative can be a place where truth is revealed without fear of retribution. We can't control our dreams, after all.

While Kepler stands firmly on the side of Copernicus and his friend Galileo, he also clearly reveres the Moon and believes it must hold some spiritual properties, Swinford told me. "He was never fully able to replace the mythical Moon with the scientific Moon," he says. Kepler was also unable to separate his spirituality and Lutheran belief in God from the worlds he saw in the telescope. Copernicus made it impossible to believe Earth was the center of the universe, but Kepler still could not remove a version of centrality from Earth's story. It is the best place, he wrote in *Dissertatio cum nuncio siderio,* or *Conversation with the Starry Messenger,* because it is the only planet from which you can see the entire solar system without any other world being obscured by the Sun's brilliance. Anyone living on Jupiter, for instance, may not see worlds like

Mercury or Earth. That is why God gave those creatures four moons instead of one for company, Kepler writes. "We humans who inhabit the Earth can with good reason (in my view) feel proud of the preeminent lodging place of our bodies, and we should be grateful to God, the creator," Kepler wrote in *Conversation with the Starry Messenger*.[5]

THOUGH HE REMAINED pious, Kepler's well-known support of Copernicus and Galileo made it harder for him to get a good job when he graduated from university. Instead of procuring a faculty position in astronomy like he wanted, Kepler was sent to practice astrology in faraway Graz, Styria, which is now part of Austria. He did not like it, scoffing that astrology is "a sortilegious monkey-play." (Shakespeare is not the only proof that the early seventeenth century was the golden age of insults.)

"A mind accustomed to mathematical deduction, when confronted with the faulty foundations [of astrology], resists a long, long time, like an obstinate mule, until compelled by beating and curses to put its foot into that dirty puddle," Kepler complained in one exemplary letter.[6]

Though he was writing *Somnium* around this time, Kepler put it aside and instead published *The Cosmographic Mystery,* another weird amalgam of mystical and evidence-based meanderings that compare Copernican and Ptolemaic cosmology. The great Tycho Brahe read it and appointed Kepler his assistant, rescuing Kepler from astrology. When Brahe died in 1601, Kepler took over Brahe's role as imperial mathematician to Holy Roman Emperor Rudolph II.

While Kepler worked on the orbit of Mars and his new laws of planetary motion, parts of *Somnium* were making their way across mainland Europe in manuscript form, before he ever formally published it. Kepler himself believed John Donne read a copy, which inspired that famous poet's satire of the Catholic Church called *Ignatius His Conclave.* Donne writes that since the Moon might be a world, all the Jesuits presently infesting this one should leave and head on up there. Though Kepler had set it aside, *Somnium* was making its way into the public consciousness.

Unfortunately, because parts of the story were autobiographical, it was easy for people to conclude that Kepler's mother, Katharina, like

Duracotus's mother, Fiolxhilde, was a witch. Kepler wrote that a first draft was carried from Prague to Leipzig, and then to Tübingen in 1611, where a local barber overheard a discussion of it and came to this conclusion.[7] Ultimately, twenty-four witnesses argued various charges against Katharina, including that she magically entered rooms through closed doors. She was formally accused of witchcraft and jailed.

THE IMPERIAL MATHEMATICIAN devoted six years to her defense. "I have no idea if this ridiculous situation, which has been blown all out of proportion, will also blow away my fifteen years of imperial service," he wrote to the Leonberg Senate on January 1, 1616.[8] "This would break my mother's heart entirely (which is, of course, my chief concern, far more important to me than any of my personal sorrows)."

Though her son finally prevailed in 1622, Katharina died shortly after her release. Her death compounded a lifetime of tragedy for the famed scientist, who had already lost his first wife and two of his children in their youth.

Somnium may have led to his mother's imprisonment, but her death only strengthened Kepler's resolve to use it to defend Copernicanism. In later footnotes to his lunar allegory, he says there are nine spirits on Levania because there are nine Greek Muses. The Levanian tutelary daemon has twenty-one characters in his name because, as Kepler explains, that is the number of letters required to spell out *Astronomia Copernicana*. The point was to demonstrate reality through a story, Kepler explains. On December 4, 1623, about a year and a half after Katharina died, he wrote to a friend named Matthias Bernegger about his goals for *Somnium*. "Would it be a great crime to paint the cyclopian morals of this period in livid colors," he said, "but for the sake of caution, to depart from the Earth with such writing and secede to the Moon?"[9] Maybe the Inquisitors wouldn't notice if his anti-Ptolemaic polemic was set on the Moon. Maybe people wouldn't be so mad if he passed it off as a wild and fantastical dream.

As further armor against criticism, Kepler even translated Plutarch's *De Facie* and Lucian's Moon satire *A True Story* on his own, and wanted to publish them alongside *Somnium* as a sort of trilogy. But he never saw a word of *Somnium* published. It only saw the light of day as a complete work—such as it was—when his son, Ludwig, published it four years after Kepler's death in 1630. Kepler had left his family penniless upon his sudden death, and Ludwig, who was in medical school, hoped the manuscript would sell enough copies to support his family.

○ •

BY THE LATE eighteenth century, the period the writer Richard Holmes has dubbed "the Age of Wonder," the Enlightenment was at full tilt. People were making new discoveries about everything from the nature of gases to the use of anesthesia. In 1781, the Americans defeated the British at the Battle of Yorktown, creating a new nation on the face of the Earth. The same year, the British astronomer William Herschel discovered the planet Uranus, expanding the boundaries of our solar system for the first time. Imagine living in such an age of upheaval—that is, if you were living through it as a literate white man who owned property, which is the only way you would have known about it. People were capable of creating new worlds! They were making new political worlds defined on paper and finding new astronomical worlds orbiting in the distant cosmos. Who knew how many other planets lurked in the great beyond? Who knew what types of beings lived on them?

THE FRENETIC PACE of change would have been terrifying, as it can be today. Waves of immigration coincided with the popularity of science fiction, where aliens were frequent stand-ins for immigrants or native inhabitants (this is still true).* And the Moon was a favorite setting for these science fiction allegories.

The Moon's gradual disappearance and return to wholeness, and its

---

* Where to even begin? Here are just a handful of selections. For allegorical treatment of ethnicity and fear of otherness, see Philip K. Dick, *Do Androids Dream of Electric Sheep?* as well as its film adaptations, both *Blade Runner* movies. For good stories about borders, citizenship, and fascism, see Robert A. Heinlein, *Starship Troopers* and *The Moon Is a Harsh Mistress*. For stories about ethics, otherness, and equal rights, see Isaac Asimov, *I, Robot*. For frank treatment of climate-driven mass migration and segregation based on race and homosexuality, read N. K. Jemisin, *The Broken Earth Trilogy*. For allegorical treatment of a big, beautiful wall dividing cultures, see Ursula K. Le Guin, *The Dispossessed*. For a sweetly allegorical story about displacement and migration, read Jill Paton Walsh, *The Green Book*, one of my all-time-favorite children's books.

innate otherness, made it a powerful symbol of change and impermanence. Just as it had been for Plutarch and Kepler, the Moon was a natural place to test out new moralities. In the eighteenth and nineteenth
centuries this sometimes meant examining the morality of enslavement
or of immigration, including westward expansion by white American
settlers who pushed into lands that had been tended, and lived in, for
centuries.

Washington Irving, the father of the American short story, offers a
shining example. Irving is best known for short stories like "Rip Van
Winkle" and "The Legend of Sleepy Hollow," the tale of schoolteacher
Ichabod Crane and the Headless Horseman. But Irving was also a prolific writer of satire. In his 1809 *A History of New York,* he imagines a race
of superior beings from the Moon who travel on mythical beasts and
decide to conquer the lesser beings of Earth, treating Earthlings much
as colonialists treated Indigenous people.

Irving sails through millennia of ideas about the history of Earth, ultimately pronouncing that it looks like an orange, bulging at the sides,
and that it's a rock. As for the Sun, he notes that the best idea of antiquity came from Anaxagoras, who maintained that the Sun was merely a
huge rock lofted skyward and ignited.

"But I give little attention to the doctrines of this philosopher, the
people of Athens having fully refuted them by banishing him from their
city; a concise mode of answering unwelcome doctrines, much resorted
to in former days," Irving writes of Anaxagoras.[10]

Irving next connects Noah (of the Ark) to the Dutch, the founders of
modern, white New York, and outlines the brutal first contacts between
European migrants and the Native American and First Nation peoples
who already lived there. He imagines what would happen if Moon citizens came down from on high and gave readers a taste of their own
medicine. The "Lunatic" leader would issue the following proclamation:

*Whereas a certain crew of Lunatics have lately discovered and taken
possession of a newly-discovered planet called the earth; and that
whereas it is inhabited by none but a race of two-legged animals that
carry their heads on their shoulders instead of under their arms;
cannot talk the Lunatic language; have two eyes instead of one; are*

destitute of tails, and of a horrible whiteness, instead of pea-green—therefore, and for a variety of other excellent reasons, they are considered incapable of possessing any property in the planet they infest, and the right and title to it are confirmed to its original discoverers. And, furthermore, the colonists who are now about to depart to the aforesaid planet are authorized and commanded to use every means to convert these infidel savages from the darkness of Christianity, and make them thorough and absolute Lunatics.[11]

Like Kepler before him, Irving uses the Moon to imply that it's better to reveal society's ignorance by refracting it through the distorted lens of imaginary others. The best science fiction makes this a goal and serves as a way to talk about otherness. Through this new lens, first polished by Kepler and later sharpened by Irving, French author Jules Verne, and others, the Moon in its nearness and realism becomes a tableau for addressing very Earth-bound problems.

Just as *Somnium* is a bridge between worlds and worldviews, Verne's *From the Earth to the Moon* offered its nineteenth-century readers a glimpse of the future. Americans are presented as overconfident, rakish land-grabbers obsessed with their guns. American political differences are portrayed as personal and vindictive. And there is even more realism in the storytelling; Verne presents the first mathematically grounded basis for space travel.

Verne is the forefather of the "hard" science fiction that infuses imaginary worlds with detailed discussions of rocketry, astrophysics, and politics. H. G. Wells, Ray Bradbury, Gene Roddenberry, Neal Stephenson, and Isaac Asimov are the inheritors of Verne's oeuvre.

From the Earth to the Moon depicts the fictional Baltimore Gun Club, who, begrudgingly at first, use their Civil War artillery expertise to build a giant cannon, the *Columbiad,* for peaceful efforts: to launch men to the Moon. Imagine a postwar society focused on finding new uses for all its guns.

Verne correctly predicts an unnerving number of things, from the launch location in central Florida to the political wrangling and the jingoism surrounding the achievement. But what sets the book apart is the particular material Verne chooses to emphasize. The highlight is not, possibly for the first time, descriptions of the Moon itself. Instead, the story focuses on the human preparation, politics, and the journey. The book sets the tone for the next century of science fiction in part for this reason. The story is propelled by the journey's technological challenges, from the *Columbiad*'s construction and launch base to the primitive telemetry used to track it.

In 1865, there was no radio, let alone live-streaming video, so the only way for humans on Earth to track the mission's progress would be through an optical telescope, just like Galileo's. The problem is, optical telescopes have to be positively enormous to see distant objects in great detail. That's because of the law of refraction, explained first by the greatly underappreciated Thomas Harriot, which dictates that how far you can see depends on the size of your glass.

To make a huge telescope worth using, you also need a clear atmosphere. The best place to find this is in a desert or on a mountain. Verne had the foresight to predict a mountaintop telescope, and he situated it on one of the most recognizable mountains in Colorado, Longs Peak.*

American ingenuity builds the telescope, but the launch is so appallingly explosive that it clouds up the atmosphere for the next several days, making Moon viewing impossible. And then, finally, the Moon emerges from the clouds, and so does the crew of the *Columbiad*—spoiler alert—not on its surface, but in Earth orbit. In Verne's telling, the Moon the visitors behold is cratered, but essentially empty. One character sees what could be the ruins of an aqueduct, but there is no life, and nothing to find.

The book is so compelling, and so clearly scientific, that it's no wonder a generation of scientists drew inspiration from it. They ultimately made its predictions real. The rockets that sent men to the Moon were originally built for warheads. And the ship that flew the first humans who landed on the Moon, the Apollo 11 command module, was called *Columbia.*

---

* To me, Verne's most egregious error is his opprobrium toward Longs and the Colorado Rockies, a range the Frenchman clearly never actually visited, or else he would not have written that "these mountains are not very high; the Alps or the Himalayas would look down upon them from their lofty heights." Actually, the Colorado Rockies contain most of North America's highest peaks, and Longs soars to 14,259 feet above sea level.

The genesis of the Moon missions of the 1960s C.E. lay in one of Verne's few, yet noteworthy, miscalculations. The travelers "had placed themselves outside of mankind by going beyond the limits which God had imposed on earthly creatures," Verne writes after they shoot off Earth through the nine-hundred-foot barrel of the *Columbiad*.[12] Reality would have been much uglier, however. The cannonball occupants would be subjected to positively horrendous g-forces at the moment of ignition, likely turning them into mush. A cannonball to the Moon just wouldn't work. Luckily for Neil Armstrong and company, a young Russian boy named Konstantin Tsiolkovsky read Verne's seminal book while in bed with scarlet fever, and became obsessed with it.

By the year that Verne died, 1905, Tsiolkovsky was an impoverished schoolteacher. Still preoccupied with the questions raised by Verne's novel, he bought books instead of food and taught himself enough physics to figure out that multistage rockets would be a better method for launching to the Moon.[13] A first stage could provide sufficient power to lift a rocket off Earth, and then a second stage could ignite afterward, giving the ship a big boost to reach escape velocity. Tsiolkovsky realized multiple rocket stages would allow humans to survive the terrific battle against gravity and discovered that pressurized gases would be more effective for propulsion than artillery, especially in space.

A decade later, the German scientist Hermann Oberth built on Tsiolkovsky's ideas, also drawing inspiration from Verne's novel, and drew up plans for multistage chemical rockets that would be safer and more efficient than giant howitzers.

A few years after Oberth worked on his plans, a young German rocket builder named Wernher von Braun—who had also read Verne and Verne's successors—saw an article in a 1920s astronomy magazine that imagined a voyage to the Moon. He would recall later that it filled him with a "romantic urge." "Interplanetary travel! Here was a task worth dedicating one's life to! Not just to stare at the Moon and the planets but to soar through the heavens and actually explore the mysterious universe! I knew how Columbus had felt," he told a *New Yorker* writer in 1950.[14] Von Braun, a Nazi and a member of the SS, had managed, like the Gun Club's president Impey Barbicane, to hastily find a lunary use for his war machines.

○ ·

VON BRAUN WAS "a master popularizer, a talented writer, and a brilliant speaker . . . a charismatic engineering manager, technological entrepreneur, and system builder," writes one biographer, the Smithsonian historian Michael J. Neufeld. And no single person did more than von Braun to advance the idea of space travel to the American public, nor to prove the practicality of space travel to American politics and industry.

Von Braun grew up in Berlin in a wealthy aristocratic family. After he made his confirmation in the Lutheran Church, like Johannes Kepler before him, von Braun's mother gifted him a telescope. He read Oberth's rocket designs as a child, ultimately working with Oberth to test liquid-fueled rockets while enrolled in the Berlin Institute of Technology. After earning his PhD in physics, von Braun toiled under the commission of Adolf Hitler. In 1940, he officially became a member of the Nazi Party and, using the labor of concentration camp prisoners, built the V-2 on Hitler's behalf.[15] The rocket, Vergeltungswaffe Zwei (Vengeance Weapon Two), Earth's first long-range ballistic missile, could carry a one-ton warhead nearly two hundred miles and killed thousands of British citizens in the Blitz.

In March 1945, with Germany in retreat, von Braun saw the writing on the wall and surrendered to the Americans. The U.S. government's secretive Operation Paperclip brought him and about a hundred other German scientists to the United States. He became the director of Marshall Space Flight Center in Huntsville, Alabama: rocket central USA. Von Braun's obsession, which he claimed was the reason for his Faustian bargain with the Nazis, was space travel.[16] He wanted to travel to the stars, and as he told later chroniclers—safely ensconced on American soil—he was less concerned with the morality of his funders and more concerned with doing the work.

After the war, space-related science fiction was all the rage. As historian Walter McDougall put it, "After V-2s and atomic bombs, any fantasy seemed credible."[17] Von Braun actively tried to channel this interest in space fiction into space reality. He published a series of vividly illustrated articles in *Collier's* magazine describing his hopes for—and the benefits of—an aggressive spaceflight program. He wanted to send artificial satellites into orbit first, followed by the first human passengers; he wanted to develop a reusable spacecraft for reliable travel to space; he wanted a permanently inhabited space station; and ultimately, he advocated for human exploration of the Moon and other planets. All of these dreams would come to pass.[18] (NASA is still working on that last part.)

Von Braun even served as technical adviser on three films Walt Disney produced in the 1950s. The second, called *Man and the Moon,* aired on December 28, 1955, and features von Braun describing a phased Moon journey, with the first stage launching from Earth and the second stage launching from a nuclear-powered space way station. The following summer, President Eisenhower—who called Disney personally and asked to borrow a copy of the first film—announced that the United States would launch a small satellite between July 1957 and December 1958.[19] Russia did this first, of course, and called it *Sputnik 1.* But von Braun would have his moment soon enough.

○ •

IF GALILEO HAD only observed the moons of Jupiter and the rings of Saturn, if he had never turned his lens toward the Moon, writers and storytellers may have imagined flights to those worlds instead. Kepler's dreamworld might have been Mars, and Verne may have envisioned a journey from Earth to Venus. Irving might have written satirical works addressing immigration that were set on Uranus, the fancy new planet. It's possible.

But even after Galileo revealed that the Moon was a world, and that other planets had satellites of their own, our Moon remained special. It maintains its mystical aura even though you can see it every day. Of course it's the first place we imagined visiting.

Lunar fiction that explored the real, scientific view of the Moon gave us the promise of lunar travel. Its proximity, and the stories that our artists and writers projected onto its beguiling face, gave us the gumption to get there.

Especially before the late twentieth century C.E., attempts to reach other planets were totally illogical. The Moon is 238,000 miles away, give or take a few miles. Mars is an average of 65 *million* miles away. The Apollo astronauts traveled to the Moon in three days; in the early 2020s, when countries from the United States to the United Arab Emirates sent spacecraft to Mars, the one-way trip took about eight months with the most powerful rockets imaginable. Mars is a long shot at best. Anywhere else is a pipe dream.

By the time von Braun wrote his popular articles and appeared in Disney movies, journeys to the Moon were no longer confined to fictional entertainment. Exploration and expansion are in our nature, as they have been since the Moon calendar makers of Warren Field planned their salmon haul, since Abraham bade farewell to Moon City of Ur and set out for the Promised Land, since my great-grandparents boarded ships and left Ireland in search of a new and better life. The Moon seemed reachable. We had the technology. It was time to try.

CHAPTER ELEVEN

The *Eagle* and the Reliquary

It is reserved for the practical genius of Americans to establish a
communication with the sidereal world. The means of arriving thither
are simple, easy, certain, infallible.

—Impey Barbicane, president of the Baltimore Gun Club,
in Jules Verne's *From the Earth to the Moon*

ANDREA MOSIE GREETED me at the doorway to the Moon with a wink.
"Are you ready?" she asked. I had been warned about the technical procedures that would precede my visit, but I was not prepared for what I
would see, and she knew it.

Only a few hundred humans have ever been to space. Only a handful
have ever walked on the Moon. But there is one other way to experience
that other realm, and it is through a doorway at NASA's Johnson Space
Center in Houston. I entered a plain-looking, boringly tan federal building, passed a wall of flags and Apollo portraits, walked up an undeco-
rated flight of corrugated stairs, and reached the vault of our stars. It is
called the Astromaterials Acquisition and Curation Office, possibly the
most interesting name ever coined for a storage room, and it is where
you can meet 842 pounds of the Moon, up close and personal. Mosie is
their primary caretaker, and she promised to introduce me.

Only five people on Earth get to touch these rocks, which comprise
the bulk of the material returned to Earth during the Apollo missions. If
you're the guy who picked up a rock from the lunar surface, they'll make
an exception, Mosie told me. But the curators who work with Mosie un-
dergo lengthy training before they begin work preparing the rocks for

scientists, who themselves spend considerable time getting ready to request just a few grams of Moon. The sample lab ships rocks across the United States and the world without any special labeling or insurance, because no amount of money can replace them. Sometimes Moon rocks are mailed out in big batches around the end of the year, when there is more holiday mail, when a heavily padded bag out of Houston might attract less attention.

Mosie is a gregarious and warm woman who has worked at the Astromaterials vault's Lunar Sample Lab since 1969, when she joined as a high school intern. Her sister Waltine Bourgeois, six years Mosie's senior, helped her get the job. Bourgeois worked at Johnson Space Center as a computer coder in the Lunar Receiving Lab, built during the Apollo missions. She punched cards to program the early computer systems. Mosie didn't have a car, so she needed to find a job somewhere she could get a ride, and Bourgeois drove her to NASA every day. At the time, she was one of the few women, let alone Black women, in her department. After her internship, Mosie wound up earning degrees in chemistry and math and working her way up through the sample lab. In 1997, she was named lab manager. It's the only place she's ever worked.

Early in her career, friends would ask if she took any of the samples home. Wouldn't it be tempting?

"No! Don't even say that! Some of these things you don't play about," Mosie told me. "Every time you touch them, you realize it's a special opportunity. It's an awesome responsibility."

Before I could enter the clean room, we had to dress head to toe in protective gear. I took off my wedding ring, my watch, and my necklace. Mosie took my clicker pen and handed me a fountain pen instead. The tiny metal springs in my clicker might shed a molecule of aluminum, she explained. No extra molecules are allowed in the clean room, lest they interfere with our measurements of the Moon. But my paper notebook was just fine. We know there are no trees up there—we know any paper dust, anything resembling anything organic, came from here, from us.

We know this now, but back in the Apollo days not everyone was so sure.

Until humans stepped on the Moon, we could only imagine what it looked and felt like. Craters made by meteorite strikes suggested the landscape was dust, but no one knew how thick it was. Before Apollo, some scientists worried the Moon would swallow spacecraft like quicksand. NASA scientist Thomas Gold worried that the dust would sink the Apollo landers, and that lunar material would be so reactive that the samples might ignite when exposed to oxygen, which is why Armstrong and Aldrin were so cautious with the omnipresent lunar dust.

Until Apollo, no one knew which elements composed the Moon. Scientists didn't know whether its craters were formed by volcanoes or asteroid impacts. They thought the Moon was monochrome, the way it appears from Earth, but the Moon visitors found its landscape full of browns, yellows, golds, even rosy pinks. People could only guess at everything the Moon had in store, which meant the first-generation Lunar Receiving Lab had to be ready for almost anything.

Several scientists even worried about protecting Earth from a lunar pandemic caused by Moon microbes. After splashing down in the Pacific Ocean, Apollo 11 crew members stayed in a quarantine unit for three weeks to ensure they did not spread any lunar germs. President Nixon waved at them through a sealed porthole. The first bits of the Moon were quarantined, too. About seven hundred grams, roughly a pound and a half, were placed in containers with mice, fish, birds, shrimp, flies, worms, bacteria, and thirty-three species of plants and seedlings. Scientists watched to make sure none developed weird Moon diseases or obvious mutations, just in case the rocks were poisonous. Everything was fine.

In the Lunar Sample Lab, a half century later, we know that we only need to worry about contaminating the Moon.

After reluctantly handing Mosie my pen, I covered my shoes in fabric booties, like the ones doctors and nurses wear in a surgical suite. I shrugged white coveralls up over my clothes, zipping the suit up to my throat. Snaps ringing my neck, wrists, and ankles tightened the suit against my body, allowing nothing in or out. I could not be allowed to pollute the Moon.

I pulled on purple nitrile gloves, donned a hairnet, tugged knee-high, astronaut-looking white shoes over my bootied feet, and turned to face Mosie. She's done this thousands of times, yet she laughed with me as I tried to shift my awkward bulk. She opened the locker room door and we stepped into an air shower. I stood on a Swiss-cheese-like rubber mat as a fan blew down from the ceiling. Any Earth dust, any human dust, on our suits was whisked away.

Finally, Mosie opened the entrance to the clean room, and I was surrounded by the Moon.

Most of it was locked behind a vault whose door came courtesy of the U.S. Federal Reserve. But the Lunar Sample Lab is an active geology lab, and bits of the Moon were laid out everywhere. The Moon was in cabinet drawers, displayed with special placards inside glass hoods, filling plastic baggies, perched unceremoniously in stainless steel baskets.

The rocks were so obviously not from around here. Most Earth rocks have a weariness, no matter where they are found. They are molded by beach waves and rain, smoothed by wind and time, covered in lichen, surrounded by trees or grass. The Moon rocks are nothing like Earth rocks. They are jagged, blocky, crystalline. Some are inky black and others are chalky, sparkling white. They look exactly like what they are, pieces of the shimmering Moon brought down to Earth. A few of the Moon bits Mosie and her team shared with me were more familiar-looking, especially the volcanic rocks, which resemble Hawaiian pumice—itself an alien-looking specimen born in the dark and molten heart of our world. Most are made up of minerals that are also found on Earth, though one called armalcolite (named for Apollo 11 astronauts Neil Armstrong, Buzz Aldrin, and Mike Collins) was found first on the Moon and is vanishingly rare on this planet. But even the rocks with familiar chemical makeup look weird and not, with apologies to Anaxagoras and Plutarch, "Earthy."

Mosie showed me some highlights, and then led me over to a microscope. The Moon people knew I had a favorite rock, and they took out troctolite 76535 before I arrived.

Apollo 17 astronaut Jack Schmitt used a rake to lift this rock off the Moon on December 13, 1972. He was walking in a valley called Taurus-Littrow, on the southeastern edge of the Moon's Sea of Serenity. Schmitt is the only geologist to visit the Moon, and he knew how to look for weird rocks. This little greenish, grayish specimen—troctolite is from the Greek *troctos*, meaning "trout," for its spotted green appearance, and *lithos*, for "rock"—has been called the most interesting sample to come back from the Moon.

It is coarse-grained, and it cooled off slowly after being forged in the Moon's depths. The most recent studies on this rock show that when it was made, the Moon had a rotating liquid core, just like Earth does. That core endowed the Moon with a magnetic field, just like Earth maintains today. But the Moon is calm now. Its magnetic field is no more. Troctolite 76535 tells this story because troctolite 76535 is the oldest rock to come home during Apollo. It is almost as old as the Moon. At 4.26 billion years of age, it is almost as old as Earth.

I wanted so badly to hold it and feel my hand pulse against it. It looked like no rock I had ever seen. It was jagged and cubelike, almost like a geode from a tourist shop in the mountains, but more alien. It had a grayish, spectral cast, and the olivine mineral within it gave the whole thing a faint hint of green. A cubical crystal grew from the top left of the rock. I kept moving the microscope dial to change the focus. Troctolite 76535 looked sharp and stiff, not crumbly at all. It was so utterly other. Up close, the greens and golds became clearer. I saw a dark flake that looked like gold or pyrite, fool's gold, and Mosie told me it was a mineral called pyroxene. The whole thing was smaller than I expected, so little and fragile inside its sterile box. The ancient, primordial Moon, almost close enough to kiss.

I thought of the white full-Moon disk kept safe by Ennigaldi, seventeen centuries after Enheduanna wrote poems to the Moon God. The daughter of Nabonidus may have believed it held something of Enheduanna's spirit. Troctolite 76535 was likewise imbued with something ethereal, something hallowed. It was the Moon *beneath* my eyes, not above them. It was touchable, instead of hanging forever unreachably far away from me. Relics like this tiny crystal are what make Apollo as special today as it was in 1969. The astronauts' postcards and the story of their adventure will live forever, but the pieces of the Moon itself are the finer legacy.

Because of Apollo, the glimmering orb that illuminates our nights and accompanies our days was finally made manifest. And from its unique, disorienting point of view, the Apollo program gave us—all of us—Earth. Viewing Earth from a distance totally revised our thinking about this planet, ecumenically and scientifically. Apollo brought the Moon down to Earth, and it brought Earth down to size. For the first time, we saw the Moon up close and Earth from a distance, and our understanding of both worlds changed forever.

○ˑ

THE FIRST CREATURES to assist in humanity's collective disorientation were not humans but tortoises, packed alongside a bunch of seeds into a Russian spacecraft called *Zond 5*. Twelve days before their launch in September 1968, the tortoises were strapped in the rocket capsule and deprived of food and water, so scientists would be able to study any changes in their bodies without being confused by the activity of their metabolisms.

The tortoises had no Earthly idea what was happening, but if they felt any sort of humanlike emotions, surely one of them was confusion. How strange, to be immobilized without anything to eat or drink for two weeks, to then be launched off the world. At least the humans who followed the tortoises knew where they were going, and had some idea of why.

Inside the capsule, the Soviet space engineers had mounted a small camera, which took pictures of Earth as *Zond 5* hurtled away from it. If the tortoises had been able to look out at our cloud-wisped world, would they have recognized it at all?

Today, you can follow a Twitter account that posts daily whole-Earth images from a satellite called the Deep Space Climate Observatory. People have been living on the space station continually since 2000. A rich guy even launched his car into the void in 2018.* Space is workaday. But until the late 1960s, no one had ever seen Earth from afar. In 1966, the short-lived *Lunar Orbiter 1* sent home a fuzzy image showing a crater-dappled, curved lunar surface, with a crescent Earth rising above it. Russian surveillance ships also sent back snapshots, including the first image of the Moon's far side. But radio transmission in the 1960s was poor, so the images were grainy, mottled, and hard to credit.

---

* Elon Musk launched his Tesla into space to demonstrate the launch capability of a new SpaceX rocket, the Falcon Heavy.

Then came the tortoise ship. *Zond 5* was the first spacecraft to circle around the Moon and come back home, carrying the raw negatives of its otherworldly postcards.

Almost fifty-six thousand miles away from home, a little more than one-fourth of the distance to the Moon, *Zond 5* snapped a seminal photo of a waxing gibbous Earth. Africa appears in the central part of the frame, the Sahara plainly visible. A cloud shaped like a cresting ocean wave swirls above the North Atlantic.* Clouds eddy around the Cape of Good Hope, and Antarctica is blanketed in white. North America is invisible, because it is not sunlit; like the Moon after first quarter, some 20 percent of Earth seems to have vanished in the void.

For the first time in the history of life, we saw Earth the way we usually see the Moon. It's easy to forget how singularly strange this was. We saw Earth as it really is: *alone.* It's so fragile, so cosmically small, so completely isolated and precious and limited. Earth is all we've got. Until the Russian tortoises and their camera silently sped around the Moon, nothing that was born here had ever left the place. So in a very real sense, it was impossible to see Earth as a world among many. We didn't have a cosmic perspective on our only home. For all our countless centuries of dreaming, we couldn't see it. No creature that breathes or swims, nothing that evolved with the tides to walk on land, no seed that bursts forth into a tree that reaches forever heavenward—nothing, ever, had left Earth behind.

The tortoises weren't the first animal astronauts; a long line of organisms, from chimpanzees to dogs to people, had preceded the tortoises into space beyond Earth's atmosphere. But the passengers of *Zond 5* were the first lunar visitors, the first beings to truly leave this planet. They were the first creatures to journey beyond Earth's orbit, to escape the sphere of Earth's influence, and to venture somewhere else.†

---

* This is technically called a Kelvin-Helmholtz cloud, and they form when the atmosphere is unstable, such as when the air above the clouds is moving more quickly than the air below them.

† Earth's gravity prevails at the Moon, which is one reason the Moon is still around. But Earth's influence is more than its gravity. Visitors to the Moon travel beyond Earth's magnetic field, which shields the planet from cosmic and solar radiation. Moon travelers go past even the Van Allen radiation belts, a region where charged particles loop around

AFTER CAPTURING THE gibbous Earth, something went wrong with *Zond 5*'s camera, and the spacecraft wasn't able to bring home photos of the Moon's surface. But it did bring back the tortoises. They might have achieved some level of reptilian notoriety had they been given names, but the Soviets just referred to them, rather coldly, as "the experimental animals." Though they were deprived of food and drink, they flew with all the trappings of a meal and the scent of home: seeds of peas, carrots, and tomatoes, wildflowers and pine; and some samples of humanity's most important crops, wheat and barley, so scientists could study how the seeds fared in space. The Sumerian beer goddess Ninkasi would have been proud.

They circled the Moon on September 18, 1968 C.E., and on September 21, their capsule splashed down in the Indian Ocean with incredible force. U.S. intelligence later reported that the capsule careened through the atmosphere with the speed and energy of a meteorite, a violent journey that would have killed a human. But the tortoises survived. Then they traveled back to Moscow, where scientists cut them open and examined every inch of their starved, desiccated bodies.

Symptoms of starvation aside, the creatures didn't seem any different from the eight control animals that stayed behind. There was no Moon cancer or rocket-jitter syndrome or any other weird, ominous side effect. The tortoises had fulfilled their mission. Their bodies reported that it must be safe out there, and safe to swing around the Moon, at least briefly.

Three months later, humans did it, too.

Apollo 8 was supposed to be an Earth-orbit mission, designed to test the command capsule and the lunar lander that would eventually take Neil Armstrong and Buzz Aldrin to the Moon's surface. But the tortoises' historic mission helped convince NASA officials to move the equipment test to the Moon instead. The point of Apollo was to beat Russian people

---

inside Earth's magnetic field. In satellite instruments and drawings, the Van Allen belts appear like a bow tie with Earth as the central knot. *Zond 5* stayed within Earth's gravitational sphere of influence, which is known as the Hill sphere and extends about four times beyond the Moon's most distant orbital locations. But by going around the Moon, and feeling its pull more strongly than Earth's, in a very real sense, *Zond 5* and its passengers were the first to truly leave Earth completely.

there, after all. So on December 24, 1968—Christmas Eve—astronauts Bill Anders, Jim Lovell, and Frank Borman became the first human beings to loop around Earth's companion.

From Earth, the Moon was a waxing crescent. The Moon was between Earth and the Sun that night, which meant its far side, the dark side for those on Earth, was fully illuminated. The three men gaped at the sight, which no human being had ever witnessed.

There were no obvious plains of basalt, the Moon's dark maria, which face Earth and give the appearance of a rabbit or a person's visage. These ancient lava fields cover some 30 percent of the near side, which is the Moon we see. They formed during violent meteor strikes, when the Moon was pummeled so hard it became drenched in its own molten rock. But they are a feature of the near side alone, owing to the Moon's formation story. The far side was tightly packed with craters and, as we know now, looks more like the dimpled face of planet Mercury, or the crater fields of Mars.

As Apollo 8 circled around to the Earth-facing near side, Anders was busy shooting photos of the surface so engineers could plan landing sites. But he took a second to look up. There, again, was the rising gibbous globe.

"Oh my God, look at that picture over there!" Anders shouted. "That's the Earth coming up. Wow, is that pretty." All three astronauts stared, and Anders and Lovell switched the camera to color film. All three would recall the sight countless times during the next five decades.

"It was the most beautiful, heart-catching sight of my life, one that sent a torrent of nostalgia, of sheer homesickness, surging through me," Borman recalled in a 1988 autobiography.[1]

The crew had a television camera on board, and they trained it on the distant Earth as they circled its Moon. Images of the lonely world shone from some one billion television sets spread across sixty-four nations. Each astronaut read from the Book of Genesis, the creation story written after Gilgamesh's Flood, and long after Enheduanna made the Moon paramount in the first literate cultures of Mesopotamia. The crew had had the Bible's first ten verses printed on fireproof pages and inserted into their flight plan.

"In the beginning," Anders read, "God created the heavens and the

Earth. The Earth was without form, and void; and darkness was upon the face of the deep. And the spirit of God moved upon the face of the waters." Tens of thousands of radio stations carried the broadcast.

BACK ON EARTH, tickers in New York's Times Square heralded the news: ASTRONAUTS BORMAN ANDERS AND LOVELL CIRCLE MOON. In April of that year, the tickers had blared the news of Martin Luther King, Jr.'s assassination. In June they told of Robert F. Kennedy's assassination. They told of the American military's losses during the Tet Offensive in Vietnam, and riots at the Democratic National Convention in Chicago, and a silent protest by Black athletes Tommie Smith and John Carlos during the Olympics, and the presidential victory of Richard Nixon. For many people, the circumnavigation of the Moon was a welcome respite from the turmoil of a singularly violent, unsettled year. Apollo 8's view of the rising Earth offered a new perspective on our fractured world. Borman later commented that Apollo 8 moved the nation forward diplomatically as much as scientifically. "It cast the country in a favorable light, at a time when there were many things that cast it in an unfavorable light," he told the Smithsonian Institution historian Teasel Muir-Harmony. NASA wanted it to be that way; the space agency and the nation needed a public-relations win. But the result was far more dramatic than even NASA officials expected.

The science fiction author Arthur C. Clarke wrote that the Apollo 8 broadcast "marks one of those rare turning points in the human history after which nothing will ever be the same again. . . . We no longer live in the world which existed before Christmas 1968."[2]

The Apollo missions were the culmination of uncountable dreams since the beginning of time, maybe since the beginning of eyesight. They were the pinnacle of human exploration, the most stunning events of the second millennium C.E. They were a new manifestation of the American spirit of technological innovation, and a very American brand of religiosity. They were simultaneously the grandest possible endorsement of the power of science, and the grandest possible example of American technological prowess. When President John F. Kennedy announced to Congress on May 25, 1961, that we'd land someone on the Moon "before this decade is out," there was no technology that could make it happen. NASA had no rockets powerful enough to get anything that far, not to mention computers small enough to travel on safe, habitable spaceships, which also did not exist. All of it was invented from thin air in less than a decade. From October 1957 to July 1971, a span of 165 months, Americans went from embarrassing space-race losers to the nation that sent human beings to the surface of the Moon, where they drove cars and hit golf balls.

○ ˙

PRESIDENT KENNEDY ANNOUNCED what would become the Apollo program about six weeks after a Russian cosmonaut named Yuri Gagarin became the first person to fly in space. It was about a month after the failed Bay of Pigs invasion, when a group of CIA-trained refugees landed on a Cuban beach and tried to topple the communist Castro government. The so-called space race was just past the starting line, and in focusing on the Moon, Kennedy was trying to shift attention away from a hemisphere in crisis. Kennedy didn't care about rocks. He didn't send brave souls to the Moon for science. The goal was to beat the Russians to the Moon, to defeat the great bogeyman of mid-century America and score a win for republican democracy over communism. The Apollo missions became the most expensive nonmilitary program ever, eclipsing even the Panama Canal construction, according to long-time NASA historian Roger D. Launius. The only comparable undertaking in American history is probably the Manhattan Project to build the atomic bomb.

Apollo was a planetary-scale boast, a giant risk with high reward, a daring gamble for a youthful nation. It was propaganda at its finest. Its astronauts were upstanding white Protestant men who had traditional families and wore casual pants and drank whiskey from highball glasses. They were red-blooded American heroes who cooked homemade pizzas and played baseball with their kids, then put on spacesuits and did the impossible. They were Manifest Destiny incarnate, bursting with promise and American masculinity, their legacy rising like the phallic rockets that launched them.

This picture forms through the blurred lens of history, through the carefully staged photographs of the astronauts and their families, and a long legacy of laudatory pop culture iconography. But the Apollo program was not a universally beloved emblem of American exceptionalism at the time. By 1962, members of Congress were decrying the increase in federal spending dedicated to the Moon shot. Public opinion wasn't much better. Throughout the 1960s, more than 40 percent of Americans opposed government-funded journeys to the Moon, according to Launius's research.[3] In 1965, Americans favored spending money on anti-poverty programs and "Medicare for the aged" more than they favored building star-spangled rockets. Social and physical scientists criticized the program as a "moondoggle," according to Launius. Opponents of the Vietnam War, figures from the feminist movement, and civil rights leaders used Apollo as an emblem for problems with the federal government.

Republicans critiqued Apollo's government-led excesses. Former president Dwight Eisenhower lamented that the Moon shot took money away from education and automation research. Meanwhile, Democrats in Congress tried and failed to redirect funding away from Apollo toward social programs, even as President Johnson argued federal spending in the South—where America's rockets are still made today—would lift up impoverished communities. He and Vice President Hubert Humphrey tried to tie the Moon shot to the larger scaffolding of the Great Society. "We can put a man on the Moon at the same time as we help to put a man on his feet," Humphrey said.[4] But visions of a spacefaring nation were clouded by civil rights struggles and by Vietnam. And many could not square Humphrey's promise to lift up people while lifting rockets. Many of those who felt excluded were not silent, even during the biggest space event of them all.

A few days before Apollo 11 lifted off, a contingent of five hundred mostly Black protesters, marshaled by the civil rights leader Ralph Abernathy, the right-hand man and successor of Martin Luther King, Jr., arrived at the gates of Kennedy Space Center. Abernathy and other members of the Southern Christian Leadership Conference brought four mules and two wooden carts, calling back to the promises of emancipation and drawing a stark contrast between the lives of working poor Americans and the gleaming white obelisk on the launchpad. They met with the NASA administrator, Thomas Paine, and they sang "We Shall Overcome."[5]

One archival photo shows Abernathy holding a sign that reads $12 A DAY TO FEED AN ASTRONAUT. WE COULD FEED A CHILD FOR $8. As Paine later recalled, Abernathy told the NASA chief that one-fifth of Americans lacked adequate food, clothing, shelter, and medical care, and that the money for the space program should be spent to "feed the hungry, clothe the naked, tend the sick, and house the shelterless." But Abernathy did not ask for a scrubbed launch; he sought a promise that NASA would task its employees with addressing hunger, and that the agency would support the movement to combat poverty and other social problems. At the protest, Abernathy thanked Paine, and told a UPI reporter he was proud of the accomplishment and believed in what Apollo 11 stood for.

"On the eve of man's noblest venture, I am profoundly moved by the nation's achievements in space and the heroism of the three men embarking for the Moon," he said. But "what we can do for space and exploration we demand that we do for starving people."

Paine later recalled telling Abernathy that NASA's goal was "child's play compared to the tremendously difficult human problems with which you and your people were concerned." He told the civil rights leader that he was a member of the NAACP and was sympathetic to their cause.

"If we could solve the problems of poverty in the United States by not pushing the button to launch men to the Moon," Paine told the group, "then we would not push that button." He gave one hundred members of Abernathy's group some VIP tickets to the launch, and suggested that NASA's space success should be used as a yardstick to measure American progress in other areas.

The morning of the launch, buses picked up Abernathy and members of his group, and they sat with other dignitaries to watch the liftoff.

◯ •

KENNEDY'S BOLD PRONOUNCEMENT in 1961 certainly gave the newly minted National Aeronautics and Space Administration a lot to do. In the span of eight years, NASA would launch the Mercury, Gemini, and Apollo programs, sending one and then two people at a time into orbit around Earth. They performed the first spacewalks and the first docking maneuvers, testing the skills and equipment they would need to fly to the Moon and return safely to Earth. While they flew, more than four hundred thousand other humans toiled below them to make sure they were safe.

Mercury, in the early 1960s, had been a success. Gemini, the middle child of the space program, was another win. By January 1967 it was time to test Apollo, beginning with the crew module that would serve as a home for the teams of three who would fly to the Moon. On January 27, 1967, Virgil "Gus" Grissom, Roger Chaffee, and Ed White climbed into the Apollo 1 capsule, called the command module, for a test run. Their actual launch was still a month away. The crew strapped in and prepared to check all their launch systems while Mission Controllers ran tests. Grissom shifted in his seat, brushing up against a bare copper wire. Voltage spiked across the wire an instant later, and the sealed capsule—pressurized with pure oxygen—erupted in flames. White tried to open the door, but the atmospheric pressure inside the cabin made it impossible. The three astronauts suffocated within seconds.

The tragedy continued to unfold over the next twenty months. The Moon race was put on hold while concurrent congressional investigations ran their course. NASA and its contractors redesigned the capsule so crew members could push the door out in three seconds, should anything similar ever happen again. The spacecraft's atmosphere was adjusted to a mixture of nitrogen and oxygen, more like Earth's own and less volatile than pure oxygen. The new Apollo command module contained fire-resistant materials and a redesigned electrical system. Armstrong would later tell his biographer Jay Barbree that Apollo 1 ended up saving lives, too. "We uncovered a whole barrel of snakes. We would have fixed them one by one," he said. "The fire forced us to shut the program down and redo it right, and we got there on the backs of Gus, Ed, and Roger."[6]

On October 11, 1968, Apollo 7 lifted off from launchpad 34 on Cape Canaveral, Florida. The crew of three stayed aloft for nearly eleven days, proving the Apollo capsule would be a safe home for even longer than was necessary for a lunar round trip. Only a few tests stood between the Moon and the first human boots to walk on it.

The lunar module, the actual lander that would ferry two men to and from the surface, would not be ready for its first checkouts for another few months. Some at NASA wanted to wait to orbit the Moon until the entire Apollo apparatus could be tested together. But NASA officials told Lyndon Johnson they wanted to go for it with just the orbiting capsule. The tortoises had already won this race; there was no reason for NASA to be slow and steady. So a few days after his successor, Nixon, was elected president, Johnson gave the go-ahead. On Christmas Eve, Apollo 8 circled the Moon carrying Borman, Lovell, and Anders with his camera. Apollo 9 rose from the launchpad in March 1969, carrying the capsule and the brand-new leggy lander into Earth orbit. Apollo 10 lifted off in May 1969 and flew to the Moon, a full dress rehearsal for the first landing.

And then finally, in July 1969, it was time. Time to leave Earth and set foot on another world. Time to make the giant leap across the void, from home to the unknown. Time to bring the Moon from imagination into reality. Time to visit the realm of the divine.

Neil Armstrong, Buzz Aldrin, and Mike Collins had trained and tested for years. In the haste to meet Kennedy's deadline, NASA launched five Apollo missions in the span of ten months, which imbued the whole enterprise with an intensity and energy that belied the care and precision of NASA's flyboys. Armstrong, Aldrin, and Collins were among the agency's best pilots. Each spent months mastering every inch of the command module and lunar lander. Aldrin focused on the science experiments, while Armstrong repeatedly tested flying in the Lunar Lander Training Vehicle, a full-size replica of the machine that he would later land on the Moon. Collins would not descend to the Moon, but would remain in command of the capsule that brought them there, and would ensure the first moonwalkers could make it home.

THE MORNING OF the launch, July 16, 1969, more than a million people packed the beaches of eastern Florida. They came in RVs and camper vans, or cars with tents or just cots. NASA alone had invited twenty thousand people, including the vice president and half of Congress.[7]

Armstrong was the first man on board, followed by Collins and Aldrin. They rode an elevator to the top of the skyscraper-sized rocket. Then they sat in the capsule and waited. It must have been interminable. Each man would tell the story countless times in the decades to follow, especially Aldrin, who never shied away from the spotlight. Still, no post hoc account can fully capture what they must have felt as they prepared for a rocket to carefully explode beneath them. *They were going to the Moon.* They were leaving Earth for that other place, for real.

We don't often pause to think about how strange the whole thing was, how strange to send a bunch of strapping young men to the Moon just because. How utterly *odd* that sentient pieces of Earth—because that's all we are, really, bits of the planet remolded by time and sunlight—made a choice to send some of their brethren away from it. What a weird thing to do, just for the purpose of seeing what it was like, and saying that it had been done. Just to do it, because it's hard, because we think we can, because we think someone we don't like might do it before us. In his autobiography of Apollo, *Carrying the Fire,* Collins himself reflected on the oddity of his space mission. Sitting in the capsule moments before launch, he thought to himself, "Here I am, a white male, age thirty-eight, height 5 feet 11 inches, weight 165 pounds, salary $17,000 per annum, resident of a Texas suburb, with black spot on my roses, state of mind unsettled, about to be shot off to the Moon. Yes, to the Moon."[8]

As the Apollo 11 crew waited, perched on their Saturn V, thousands of people working in Florida and in Houston checked screens and ticked boxes. They each reported "GO" as they checked and rechecked the status from every handwoven computer wire, every carefully molded rubber hose, every solid bit of fuel. *GO.*

Apollo 11 was ready. Humanity was ready. "The time for farewells had come."[9] At 9:32 in the morning Florida time, on July 16, 1969 C.E., the Saturn V rocket ignited. The tower strained against the gravity of Earth, the planet that birthed us, the only planet covered in life, the only one accompanied by a solitary spectral Moon. The rocket trembled. Armstrong, Aldrin, and Collins felt themselves smooshed against their seats as the rocket strained against the atmosphere. Bolts fired and released the Saturn V from its shackles. Then it did the improbable, and the in-

exorable: It left. Promethean thunder scattered the birds and a shock-
wave rippled across the beach, ruffled the ties of the assembled men,
swished the updos of the women, pressed against the hopeful chests of
children who didn't yet know space launches could end in grief. In a moment, the rocket and its flame were a speck in the sky as they sped
Moonward. The astronauts' target was 218,096 miles away.

$$\bigcirc \cdot$$

THREE DAYS PASSED. Earth grew smaller in the window and the crew's
quarry grew larger. The astronauts named their command ship *Columbia.* It was not at all a coincidence that the ship lofted by the brains of
von Braun would carry the moniker of Jules Verne's Moon ship. They
named the lunar lander *Eagle.* Their emblem was our national mascot, alighting on the Moon carrying a small golden replica of an olive
branch, a global symbol of peace. The eagle comes to us from the Roman
legions, and the olive branch comes down to us from the story of Genesis. But like that story, the bird has an even older heritage. After the
Flood, Noah sends out a dove, which returns bearing an olive branch as
a sign of dry land and a truce with God. In the *Epic of Gilgamesh,* after
the Flood, Utnapishtim sends a dove, a swallow, and a raven. Only the
raven doesn't come back, and Utnapishtim knows the bird stayed behind, having found what he sought.

The Apollo crew arrived in lunar orbit on July 19, and the next afternoon, Armstrong and Aldrin floated into the *Eagle* lander, closing the
door of *Columbia.* Collins orbited the Moon while the pair descended.
On the afternoon of July 20, they undocked from *Columbia* and turned
on the *Eagle*'s engine. The ship began to descend toward the lunar surface. Armstrong and Aldrin felt their bodies grow heavier again, though
faintly, as the Moon's gravity reached toward them. They were falling
now.

A few hundred feet from the surface, Armstrong took command of
the *Eagle.* He looked out the window. He and Aldrin had scrutinized the
photos made by Apollo 8 and 10, and they knew where they wanted to
land. Armstrong watched the target crater recede in the distance. They
were almost four miles off the mark.

Armstrong, by many accounts the most careful pilot in the astronaut corps, had the controls. His hands tensed and relaxed with surgical precision, puffing gas through the *Eagle*'s thrusters to adjust its course. Nothing but boulders lay beneath them. If Armstrong set the lander down on a slope or a rock, the *Eagle*'s legs might tip over, meaning it would never take off again, consigning the two astronauts to the Moon forever. He scanned the horizon as Aldrin called out their altitude above the Moon. They were falling at twenty feet per second.

"120 feet." Aldrin's voice was tranquil. Mission Control was silent.

"Okay. 75 feet. There's looking good," Aldrin noticed where Armstrong planned to set down.

Houston beeped in. "Sixty seconds," said Charlie Duke, who was serving as CAPCOM, an astronaut on Earth tasked with crew communication. One minute of fuel remained. If it ran out before the *Eagle* touched down, they would crash.

They fell further. Thirty seconds. Silence from the lander.

"We copy you down, Eagle," Charlie Duke said. Hundreds of millions of people on Earth held their breath.

"Houston, Tranquility Base here. The *Eagle* has landed," Armstrong said evenly. Duke was less calm: "You got a bunch of guys about to turn blue. We're breathing again. Thanks a lot," he said.[10]

"Thank you," Armstrong said. *Thank you. Yes. I am now on the Moon. Thanks.* The poise in his voice on the recordings is hard to comprehend, an engineer's calm and a pilot's formality. How many other people would have let out a barbaric yawp? At the very least, a gasp?

NASA told the media that Armstrong and Aldrin would go to sleep for four hours, which did not happen; instead, the pair readied themselves to get out. By the time they were ready, it was time for Sunday dinner in America, and early morning in Europe. Schoolchildren around the world were given July 21 off. One-fifth of humanity watched on live TV, in an era when the vast majority of people didn't have televisions. Mission Controllers in Houston told Collins, orbiting above in *Columbia*, that Armstrong and Aldrin wanted to get out early. "Tell them to eat some lunch before they do," he said.[11]

As he prepared to exit the lander, Armstrong realized his piloting prowess had almost gotten them into trouble. The *Eagle* was supposed to touch down with a bit of force, which would collapse its legs a bit and make the descent down the ladder a little easier. Armstrong alighted like a hummingbird, the barest touch bringing the *Eagle* down. That meant the lander's spiderlike body hovered far above the surface, making his first small step more like a jump. The last rung of the ladder hung three feet from the Moon's surface.

He opened the hatch.

He got out.

He checked the distance to the ground—"it takes a pretty good little jump," he warned Aldrin—and prepared to walk on the Moon.

"I'm going to step off the LM now," he said. And at 10:56 P.M. Eastern Daylight Time, on Sunday evening, July 20, 1969 C.E., he did. You know what he said next.

NEIL ARMSTRONG WAS the only man on the Moon for about eighteen minutes. He walked around a bit, getting his bearings, and accidentally trampling his first bootprints. He'd meant to take a photo of those. He scooped a quick sample, in case he and Aldrin had to evacuate suddenly. And then he photographed Aldrin coming down the ladder. The photo sequence shows Aldrin's leg out, arm holding on, descending, in one of the most active photos from the duo's brief jaunt on the Moon. This photo, more than many of the others, brims with meaning.

Until Armstrong photographed Aldrin, every image of the Moon had been from afar—from a distance, looking up, or even looking down from its orbital embrace. But now, humans were on it, on the inside looking out. The influential twentieth-century photographer Edward Steichen once noted that "a portrait is not made in the camera but on either side of it." Once captured, the moment really exists among three parties: the subject, the photographer, and the viewer. In this photo of Aldrin climbing down, all three are on the Moon for the first time. The human mind was *inhabiting* the Moon for the first time, instead of projecting onto it from the outside.

The landscape was unlike anywhere on Earth. It had that acrid gunpowder aroma. Its surface seemed to undulate, a trick of the terrain and the stark shadows cast by angled sunlight falling flat, with no atmosphere to bend it. The surface was sandy and soft, which Aldrin later compared to finely powdered carbon.[12] The Moon was not gray, but a landscape flecked with color. Apollo 10 astronauts sailing above its surface had noticed a range of browns, from light tan to milk chocolate. Buzz Aldrin remarked at one point that some rocks resembled biotite, a greenish-brownish-black substance. He joked that some rocks near the surface even seemed to sparkle with purple.[13] Analysis of the Moon rocks has since found volcanic glass in every color of the spectrum, from yellow and orange to green and brown and blue. Despite the vivid surroundings, the landscape was deathly quiet. The astronauts talked to each other through voice-activated microphones, so their helmets were full of machine beeps and garbled commentary from Earth and each other. But if they could have listened to the Moon, they would have heard nothing at all.

Though Armstrong's famous first step and the precious rocks the pair collected will live forever, their lunar sojourn lasted only two hours. After all that, they were supposed to sleep.

They tried to rest, but whirring life-support pumps and bright sunlight foiled any attempt at real shut-eye. Later Apollo missions carried hammocks so the astronauts could sleep a little more comfortably, but Armstrong and Aldrin had to curl up on the *Eagle*'s floor. Armstrong's nose was stuffed up, probably because of all the Moon dust they had tracked inside, and because of space itself. Many astronauts report feeling congested in low gravity. It was also too bright to sleep. The pair pulled shades over the lander's windows, but the *Eagle*'s instrument panels were illuminated, too. "After I got into my sleep stage and all settled down, I realized there was something else," Armstrong said at a later debriefing. He was nestled beneath the *Eagle*'s telescope, which the crew used to chart their journey against the stars. "Earth was shining right through the telescope into my eye. It was like a light bulb."[14]

The first human being to step on a world besides Earth blocked his home planet's light with a rock-collecting bag.

○ •

WHEN NEIL AND BUZZ and Mike came home, the world wanted a piece of them. The Nixon administration sent them on a tour, including their wives and a few NASA officials and a secretary named Geneva "Gennie"

Barnes. They visited twenty-seven cities within twenty-four countries in a dizzying span of forty-five days.

The astronauts knew their journeys were monumental, but in the context of the 1960s, and in the context of their daunting engineering and piloting work, they put poetry aside. Collins, who stayed alone in the *Columbia* orbiter while his colleagues went gallivanting about, once lamented that the astronauts were not equipped to talk about what they witnessed.

"We weren't trained to emote, we were trained to repress emotions, lest they interfere with our very complicated, delicate, and one-chance-only duties," he demurred in his memoir. "If they wanted an emotional press conference, for Christ's sake, they should have put together an Apollo crew of a philosopher, a priest, and a poet—not three test pilots."[15]

The immortal tale became a burden to many astronauts. Neil Armstrong came home as the most famous person alive and largely retreated from public life. The other Apollo moonwalkers often struggled to re-enter the regular world—the one we all live on, the quotidian one with meetings and humdrum tasks to which the rest of humanity has eternally been consigned—after witnessing that world recede in the distance. Several of them got divorced after their return. Buzz Aldrin has written at length about his alcoholism. Jim Irwin, who collected some of the most important rocks to come home from the Moon, became a born-again Christian and dedicated himself to spreading the Gospel, once saying "God walking on the Earth is more important than man walking on the Moon."[16]

Some astronauts tried to translate their un-Earthly experiences in other ways. Russian cosmonaut Alexei Leonov* and Apollo 12 astronaut Alan Bean both produced art after returning from space. Bean retired from NASA and became a full-time painter after returning home as the

---

* Leonov was one of the Russian space program's most accomplished cosmonauts and had hoped to become the first person on the Moon. He lost to the Americans, but he was the first person to conduct a spacewalk, exiting his Voskhod 2 capsule for about twelve minutes in 1965.

fourth man to walk on the Moon. His canvases are textured with boot-prints and crater dimples. The paintings are wild with color, streaked with golds and pinks, purples and oranges—colors that astronauts and the rocks report were true, but that look nothing like the stark black-and-white images captured by the cameras. Many astronauts had described the vivid colors of the Moon, but Bean was the only one to dedicate his remaining years to rendering them for the rest of us.

The astronauts were human, so naturally they tried to have fun while ticking items off their endless checklists. Walking on the Moon was a joyful exercise; they bounced! They had a hard time getting back up when they fell over. They reported that the views were magnificent, spectral, colorful. The technical, jargon-filled Apollo transcripts are peppered with moments of delight.

But even their joy is tempered with a certain kind of grandeur. The Moon was strange, walking and working on it was strange, and as Irwin put it, it changes you. Moonwalking was revelry limned with ecstasy, of the kind that would be familiar to the faithful.

Many astronauts were pious people. Of the thirty-two men assigned to the Apollo missions, twenty-six were Protestant and six were Catholic. They served as leaders in their congregations, a narrative that neatly fit NASA's picturesque Americana propaganda. Several of the men spoke of their faith in the Christian God, the God of Abraham of Ur, Moon City of the ancient Near East. They did so even after a lawsuit from an American atheist named Madalyn Murray O'Hair, who sued the space agency for First Amendment violations after the Apollo 8 Genesis reading. Aldrin, a Presbyterian, worked out an arrangement with his church to quietly consume a thimbleful of wine and a Communion wafer while on the lunar surface. In a ceremony that took place during radio blackout, he read from the Gospel of John: "I am the vine, you are the branches. Whoever abides in me, and I in him, will bring forth much fruit. Apart from me you can do nothing."[17]

And later, en route home from the historic first landing, Aldrin took part in a TV broadcast during which he quoted from Psalms: "Personally, reflecting [on] the events of the past several days, a verse from Psalms comes to mind to me. When I consider thy heavens, the work of thy fingers, the moon and the stars, which thou has ordained; What is man that thou art mindful of him?"[18]

From Nabonidus and Sin to Copernicus's papal dedication, astronauts like Aldrin, Irwin, and the Apollo 8 crew followed a long tradition of explorers invoking a higher power in command of their cosmic pursuits. For the space explorers, some of this was demographically predestined. America in the 1960s and 1970s was a more overtly religious nation than it is now.[19] Among people born in the "Silent Generation," from 1928 to 1945, a whopping 84 percent of Americans identified as Christian, according to the Pew Research Center. This generation includes every member of the Apollo astronaut corps. But elements of the astronauts' religiosity were also baked into their mission.

Project Apollo was, metaphorically and truly, a mission much like an organized religion. It grasped for immortality. Its political goals, set against the backdrop of the Cold War and the civil rights movement, were in some sense a way of seeking absolution. Its messengers were exemplars of courage and conviction, and their tasks inspired a kind of reverence. The missions even brought about a new awakening, in this case new knowledge and a different way of thinking about humanity's home and our shared experience. In the simplest terms, Apollo was about leaving Earth behind, a mission that until 1968 had been limited to stories about gods and higher beings. It makes sense that the people who actually transcended Earth were imbued with mythical power.

There is another reason for the religion-soaked commentary and ritual. Regardless of their faith traditions, most people who leave Earth have borne witness to the journey's transformational power.

By the early 2020s, only about 550 human beings had slipped the surly bonds of Earth and floated above it far enough to see its curvature, to see the eggshell-thin line of our life-giving atmosphere. Many report feeling an overwhelming sense of clarity and unity, a heart-swelling state of heightened awareness and togetherness that is common enough to have its own name: the "overview effect."

The term comes from aviation, and if you've ever flown in a plane you might have experienced a version of it. The sense of boundaries evaporating, of your own puniness when arrayed against the broader world;

the unnerving shift in perspective as cars turn into tiny dots and metropolises melt until you notice only land and water. You realize we are
all hanging on to the same rock, the only place any of us can turn in
deadly turbulent waters. You realize it's worth protecting. You realize its
inhabitants are worth protecting.

This is how Edgar Mitchell, the sixth man to walk on the Moon, described the overview effect in 1974, in an interview with *People* magazine:

> *You develop an instant global consciousness, a people orientation, an*
> *intense dissatisfaction with the state of the world, and a compulsion*
> *to do something about it. From out there on the Moon, international*
> *politics look so petty. You want to grab a politician by the scruff of the*
> *neck and drag him a quarter of a million miles out and say, "Look at*
> *that, you son of a bitch."*[20]

A visitor to space does not need to be pious to feel this epiphany.

○˙

BACK ON EARTH, in Houston's Lunar Sample Lab, it was possible to experience a similar version of this otherworldly awe. After Andrea Mosie
took me through the first door, we performed a series of steps that felt
very much like a purification ritual. I removed any Earthly pollutants
from my person. I clothed myself in pure white, handling each step with
care and something like fear, because I didn't want to contaminate the
special garments. I walked through a literal cleansing shower before I
could approach the Moon. As someone who grew up with the rituals of
Catholicism, it was a bit familiar. I felt separated from common things
and closer to spirituality, similar to how I felt inside the stone circles of
northern Scotland. The place and the process were designed to be special, and so the experience was special, too.

The sample lab's attention to lunar detail also imbues the place with
something almost holy. Before the astronauts lifted the rocks, they photographed their orientation on the Moon and noted how the rocks came
to rest. The Lunar Sample Lab keeps that orientation, so that the rock is

stored according to the way it was found. That way, scientists who study a sliver of it will know whether that side was exposed to the void and bombarded by cosmic rays, or whether it was protected by the Moon. All this information provides a baseline for scrutinizing the rock's makeup, but it also endows the samples with a bit of mysticism. *This is exactly what it looked like, up there,* the orientation says. *You, too, can glimpse the spotless relic. See where it takes you.*

The sample preparers stick their covered arms through rubber gloves attached to a cabinet filled with pressurized air. They employ forceps to pick up sheets of Teflon, which they use to wrap the Moon bits before stuffing them into plastic baggies. These are sealed before the rocks enter Earth's atmosphere. Then they are placed in small canisters for shipping. Mosie's team takes extreme care to ensure the Moon material does not touch Earth or any of its creations until scientists are ready to receive their samples. The work is tedious and artful; it's tricky to manipulate tiny shards of rock while wearing three layers of rubber gloves. But the curators are not clumsy. They treat the rocks with a kind of reverence.

The Lunar Sample Lab is no Holy Sepulchre, of course, and the Moon rocks are not religious relics. But visiting them is a sort of devotional pilgrimage nonetheless, a Houston Hajj to a different kind of Black Stone. These rocks tell the story of our origins. They tell of a very real and very ancient Flood, one that covered the Moon, one that covered Earth from which the Moon was made.

TOWARD THE END of Neil Armstrong's short lunar sojourn, he visually weighed the treasure chest and found it wanting. He dumped in nine extra shovelfuls of sand to fill it up. He could never have expected that the last-second scoops contained some of the most important rocks to ever come down to Earth. "Neil turned out to be the best field geologist on the Moon, until Apollo 17, that is," Jack Schmitt, a field geologist who walked on the Moon during Apollo 17, told me in 2019. "In twenty minutes, he collected some of the most interesting samples we got. Had we never had another mission, we would have gotten enough."

Armstrong's afterthought rocks were anorthosite, that low-density material that forms when minerals crystallize within molten rock. Their existence hinted that the Moon was once bathed in an ocean of magma. "We knew right there that the Moon would be a record of the early history of the Earth," Schmitt explained to me. "That was not clearly understood before Apollo 11, but it was understood afterward, and now."

After Apollo 11's triumph, NASA's confidence grew and later missions lasted many days. The last three missions even sent a car so astronauts could drive farther and collect more unusual rocks. Apollo 15 astronauts Dave Scott and Jim Irwin, the one who later dedicated himself to Christ, went to the Moon with a wish list. Geologists were hoping for volcanic rocks, glass beads that form inside fire fountains, and other samples that would shed light on the Moon's youth. On their second day on the Moon, the Apollo 15 moonwalkers struck gold. They were taking samples of material ejected from a crater called Spur, inside the Moon's Mare Imbrium, or Sea of Rains, when Scott saw an unusual rock, which sat on what he would later describe as a pedestal of soil. He lifted it and dusted it off, and the astronauts recognized right away that the rock was special.

"Oh, man!" Irwin shouted. "Look at the glint! . . . Almost see twinning in there!"

Both astronauts started whooping with glee.

"Guess what we just found," Scott radioed Houston, as Irwin laughed. "Guess what we just found! I think we found what we came for." Both men hollered and started describing the rock. "What a beaut!" Scott said.[21]

It was an anorthosite, and a big one. Reporters covering the mission were told it could help tell the story of the Moon's formation from a vast flood of oozing rock. They named it Genesis Rock.

Genesis Rock was sitting in a pressurized box the day I visited. It was so very, very white. It was in a plastic baggie, and some crumbs filled the bag's right corner: little bits of the Moon that had started to flake off and fall apart, as prosaic as the dust from the bottom of a cereal box. The crumbs were the perfect encapsulation of the new worlds Apollo brought to us: rocky, maybe even "Earthy," three-dimensional, tangible, real.

THE MOON SHAPED our evolution and served as our timekeeper and our spiritual lodestar through the ages. It guided the grand barque of civilization through the dawn of religion, the onset of philosophy, the age of exploration, the ideals of the Enlightenment. But only in the age of Apollo did it finally exist as a material place in space and time.

The Apollo missions were designed to use the Moon as a tool. It was an instrument of might, just as surely as it was for the stone circles of northern Scotland, the Nebra sky disk, and the temples dedicated to Sin. Americans walked up there to show they could do it, and in doing so, demonstrated what glory was possible through democratic republicanism and white Protestant Christianity, rather than Soviet communism and godlessness.

But the Moon during Apollo also gave us something unexpected. Like the work of the sky priests toiling under Nabonidus, Apollo's expeditions bestowed upon us something we weren't necessarily seeking. Apollo was a civic errand, but our journeys to the Moon gave us new knowledge about the universe and our place in it. The Moon reoriented our ideas about that universe. The heroic journey came first, steeped in myth like the Babylonian sky records. New knowledge, a new logos, came after.

Because of the Apollo missions, the Moon is no longer a faceless luminary in our imaginations; it isn't just a dreamworld where brave characters go exploring, whether in novels or on real, live TV narrated by Walter Cronkite. Through Apollo, the Moon became an awesome three-dimensional sphere. Its physicality became measurable and quantifiable. The "terribility of her isolated dominant implacable resplendent propinquity," as James Joyce described the Moon, was literally close at hand. The Moon's surface was well-trodden now. We sifted it in our very real hands, its dead soils slipping through fingers pulsing with life. Thanks to Apollo, it is here, among us. A bit of Earth returned triumphant, our patrimony brought home, to tell us all our own story.

CHAPTER TWELVE

Our Eighth Continent

MY DAUGHTER'S TINY purple hiking boots crunched in the dirt as we climbed up the steep trail, ascending stone steps and exposed rock slabs. Juniper and scrubby pines bordered the trail to our left, heading toward Colorado's San Juan mountain range. Nothing guarded the cliff on the other side, and the road was visible a vertiginous drop below. When I turned around, I could see the valley floor, opening toward the barren landscape of northern New Mexico. In front of us loomed the geologic feature we were there to see: two towering spires of sandstone, with a U-shaped gap between them.

One hundred million years ago, the towers made up the sandy floor of a vast interior sea. When volcanoes lifted the Colorado Plateau forty million years ago, the dry seafloor was exposed. Eons of wind, rain, and water flowing through the Piedra River etched away the sandstone, leaving two towers, one known as Chimney Rock and one uncreatively known as Companion Rock. They tower above a village of ghosts.

One thousand years ago, the spires soared above a bustling settlement populated by Anasazi, a people now more commonly known as the Ancestral Puebloans. They built more than two hundred structures here and lived in them for many years. Archaeological excavations have found wooden beams that were erected in 1076 C.E. and others that were built in 1093 C.E.—almost eighteen years apart. People lived here for several decades after that, and then, suddenly, they left and never returned.

Anasazi dwellings are found to this day in the mountains throughout my home state of Colorado and surrounding states, carved into the rocks, sculpted from the hills that protect them, in crevasses that guard them like eaves. They lived in Mesa Verde, Chimney Rock, and Chaco Canyon. They disappeared before the arrival of European settlers who would come to dominate America.

The Ancestral Puebloans lived around 800 C.E., but their rhythms recalled much older civilizations that rose and fell in places like Aberdeenshire, in Scotland; like Jericho, in Palestine; and near Nasiriyah, in Iraq, where the Moon City of Ur was located. The Anasazi used the Moon and the stars to chart their lives, to learn and follow the seasons, and to plan for the year. Near what is now Pagosa Springs, Colorado, they constructed a house that turns its face to the Moon every 18.6 years.

Thirteen years before I visited the spires with my older daughter, the Moon rose at sunset above the deserted village and its stone houses. The first shimmer of its light crested the horizon between the two spires, and from the house my daughter and I were heading toward, the towers seemed to cradle it like a hug.

The place is now called Chimney Rock National Monument. It includes several stone buildings erected by the Ancestral Puebloans, but most important is the Great House, a stone building with a great round room at its center, called a kiva. The house was erected in the style of Chacoan Great Houses, which can be found peppered throughout the American Southwest. The Ancestral Puebloans built these houses with straight walls, air ducts, ventilation shafts, and core walls insulated with veneer walls to give the buildings strength.

Drought and changing cultural patterns likely dispersed the Anasazi over more than a century. But Chimney Rock was abandoned abruptly between 1125 and 1130 C.E., earlier than other Anasazi sites, earlier than all other Ancestral Puebloan sites except Chaco, in what is now New Mexico.

THE PEOPLE WHO built this structure were not thinking of convenience. It's far from any water source; the Piedra River is a thousand feet below the Great House. It looms far above the fields where they would have grown maize, beans, and other crops. It was not built for easy living.

The Ancestral Puebloans often lifted their houses to the skies. But the Chimney Rock Great House is even loftier and was probably built for ritual. It was erected in exactly the right spot to see the two towers embrace the rising Moon.

Remember the stone circles of northeastern Scotland, in Easter Aquhorthies and Tomnaverie, and their arrangement in honor of this roughly nineteen-year round? Owing to the movements of the celestial bodies, the Moon returns to roughly the same patch of sky after twelve lunar cycles. But it does not return to the exact same spot. It rises at different points on the horizon over the years, just as the Sun changes where it rises and sets on the horizon during the course of one year. The cycle of rising and setting is complete after 18.6 years. At the end of each arc across the horizon, the Moon appears to pause in place for about three years, rising at the same point on the horizon before starting to move in the opposite direction on the horizon each night. This long phase is known as the major lunar standstill.

In July of these years, the whisper-thin new Moon rises between the spires at Chimney Rock just before dawn. At the winter solstice in December, the full Moon rises between the spires at sunset. During the spring equinox the following year, a half-Moon is visible through the monoliths at midnight. The summer solstice new Moon rises in the rock cradle at sunrise. The Ancestral Puebloans set one of their stone houses into the side of the cliff for the best possible view of these events.

Were they setting their calendar, like the pit diggers of Warren Field and the metalsmiths of the Nebra sky disk? The Ancestral Puebloans were agrarian people, and it's possible they wanted to keep track of the seasons with a year that began at the winter solstice. It's possible the Chacoan people traded or connected somehow with the Maya or Inca of Central America, who also observed these lunar phenomena and their relation to the solstices. The Sun also rises through the pillars in the spring, so maybe they used the rock formation to study both celestial lights. Maybe the Puebloans who built Chimney Rock's houses used the natural towers for ceremonies we no longer know or understand, ones we assume were ritual in nature.

Archaeological evidence shows that the site hosted large fires, probably intentional bonfires, likely set as signal fires. I thought immediately of Aberdeen, where the people who gathered under the full Moon at its standstill, every 18.6 years, lit bonfires to communicate with others. Standing on the ridge in Colorado, I imagined a succession of small fires dotting the landscape below us, down into the valley, down into the canyons of New Mexico. Watchmen set on neighboring ridges might have lit the next blaze in a signal relay. Flame and smoke are visible for miles here—something the Forest Service can attest to. The Civilian Conservation Corps erected a fire detection tower beside the Great House in 1940. It was removed in 2010 so future lunar standstills would be visible from the site.

Maybe these people were trying to communicate with their kin, people living in places like Pueblo Alto, down the valleys into New Mexico, down to the great town of Pueblo Bonito within Chaco Canyon. Maybe they were communicating with their relatives and neighboring clans. Maybe, like at Tomnaverie, they were communicating with another realm entirely. I wondered about this other realm as I walked with my daughter, and then I came across this Tewa Pueblo song printed on a trail marker:

> Yonder comes the dawn
> The universe grows green
> The road to the underworld is open!

At the height of Chimney Rock's settlement, some two thousand people lived in eight villages spread out across a five-mile radius. They grew squash, corn, and beans, and collected seeds and other plants for food and medicine. Some of the storerooms around the settlement contained pottery painted with spiral designs, big flasks for storing water or other drinks, carved animal figurines, and flat stones with holes bored into them, maybe used to hold candles or incense. Archaeologists from the University of Denver who excavated Chimney Rock in the 1920s and the 1970s found collections of high-quality pottery, seemingly too lovely to leave behind. The beautiful ruins are among the pieces of evidence suggesting the settlement was abandoned in haste.

In some rooms, the people who lived here kept materials we also don't understand, so we assume they were used for rituals. Other rooms were empty when archaeologists excavated them in the 1970s. They might have been guest rooms, used for certain times of year when people visited from more distant reaches, like Chaco Canyon and other Chacoan communities.

Throughout the Southwest, archaeoastronomers have found evidence that Ancestral Puebloans and their contemporaries around North and Central America possessed detailed knowledge of astronomy. In many places petroglyphs carved into boulders are arranged to catch sunlight or moonlight at certain times of year. In Chaco Canyon, on a rock slab on a landform called Fajada Butte, Chacoan people carved a spiral that catches a triangle of sunlight on the summer solstice. During the lunar standstill, a shadow crosses the same spiral when the Moon rises.

IMAGINE FIGURING OUT these patterns ten centuries ago, with no written language, no known system of written mathematics, no formal calendar that survives to now. There were no astronomical charts, no telescopes, not even cuneiform tablets. But at some point, an ancestor figured out how to bring the Sun and Moon to Earth, how to catch their light, and how to use that light to mark time. This person took a chisel, hiked up to a crevice on the butte, and hammered the right boulder in just the right place. Their contemporary in Chimney Rock looked up, witnessed Earth embracing the rising Moon, and marked the spot: Here is where we should build our Great House, here is where we will hold our calendar ceremony, here is where we will make our lives.

The Moon gave the Chacoan priests, like the calendar makers of Warren Field, control of time. In turn, this ability gave them control of society, in the same way as the artisans who made the Nebra sky disk. Management of the calendar meant jurisdiction over ceremonial rituals, trade festivals, harvest times and feasts, and even social events. Like the sky priests of Nabonidus, the astronomers of the Chacoan world would have led lives of comfort and prestige. They might have worn elaborate clothing, turquoise beads and macaw feathers on their heads, maybe special headgear like the Shamaness of Bad Dürrenberg. They might have played drums or flutes, might have sang tunes around a hearth like I sang with Charlie Murray.

The Ancestral Puebloans used the Moon in ways we tend to describe as ritualistic or spiritual, but this characterization is likely born of our ignorance. When we don't know how to communicate with a vanished people we describe them as mysterious. But there is no difference, really, between their vision of the Moon and my own. There is no difference between the power they derived from the Moon and the power imbued in the Nebra sky disk, or the Berlin Gold Hat, or the telescopes of Galileo and Harriot. There is no difference, fundamentally, between the Moon in the eyes of the European explorers of the Americas, who followed the Moon over the Atlantic and dreamt of riches, and the Moon in the eyes of Amazon CEO Jeff Bezos, who sees its surface and thinks of riches. There is no new thing under the Moon, to paraphrase a scribe from Mesopotamia.[1]

The Moon can bring us knowledge, pure information, for its own sake. And it can also afford us a spiritual experience. It helps us understand time, both the daily rhythms of human life and, from its quiet vantage, the grander scale of the entire universe, to the time before we evolved, to a time before even the first stars existed. The Moon still gives us everything it has ever given us. It reflects what we want it to reflect, in our particular culture, in our particular time.

So what do we owe it in return?

○ ˙

WHEN NEIL ARMSTRONG and Buzz Aldrin departed the Moon in 1969, they left a commemorative plaque mounted on the legs of their *Eagle* lander. It contained the Apollo 11 crew's names and signatures, plus that of the American president at the time, Richard Nixon. It contained a simple projection of Earth's two hemispheres, showing all the landmasses according to their plate tectonic arrangement as of 1969 C.E. And engraved in a rounded, 1960s-style font were the words:

HERE MEN FROM THE PLANET EARTH
FIRST SET FOOT UPON THE MOON
JULY 1969, A.D.
WE CAME IN PEACE FOR ALL MANKIND

The Moon landing occurred during, and because of, the Cold War. So in one sense it is impressive that the language included "for all mankind" in that (unfortunately gendered) last line. NASA officials didn't need to say this. They could have included a silver-and-black engraved American flag instead. But they acknowledged the world, because even in the height of the space race, they recognized that if the Moon belongs to anyone, it belongs equally to everyone on planet Earth.

We might never have stared at the Moon's peaks and valleys in such detail if Mesopotamian artisans had not figured out how to mix sand, soda, and lime to make glass. The NASA "computer" Katherine Johnson's calculations for the Apollo 11 trajectory rested on the mathematics of the Greeks and Babylonians who learned numbers in the measurement of time. Apollo may have been an American effort (with the help of some German expats), but the achievement was the whole world's.

Yet in the early twenty-first century C.E., the Moon is widely perceived as a place to build, to extract, to maybe get rich or die trying.

Some modern scientists now argue we should use the Moon to move some types of dirty manufacturing off this fragile planet and onto its barren companion. As people begin realizing that Earth is in fact special, and that we have no second option, more humans (if not those in power) have started arguing more vociferously for ways to protect Earth and its resources. But the Moon is not quite as valuable, according to this argument. Aristotle's sublunary realm is now the perfect one, and the heavens are ripe for exploitation.

Humans have touched the Moon—*Americans* have touched it—but that is not enough for China, or India, or even the United States in the twenty-first century C.E. The leaders of these countries each want their particular brands of boots to mar its surface, and their own particular twenty-first-century names engraved on plaques to be mounted forever in its dust. We have to identify the real reasons why we'd like to revisit the Moon, and what we would do once we got there. Will we go back because it's still hard? Because we have something to prove, some faceless enemy nation to impress? Because we would like some rocks in the name of science? Because we would like to make money? Or because of something grander, something more ineffable?

From Kepler's *Somnium* to von Braun's dreams to NASA's new Artemis missions, which seek to return Americans to the Moon in the 2020s, our celestial companion cannot be rinsed of myth. The Moon has always been spiritually special, even when it was a tool for power. The drumbeats reverberating off the white stone of Tomnaverie, the chants that accompanied the use of the Nebra sky disk, the pious Moon prayers of Nabonidus, the now-unknown ceremonies hosted in the Chimney Rock Great House—none of these rituals could be erased from our journey with the Moon through time, not even after Galileo unmasked its true nature. Mysticism still imbues our visions of the Moon. Today, spaceflight offers a form of transcendence, and is almost a form of civil religion in modern America. Modern space exploration is limned—sometimes explicitly, depending on who's talking—with the idea of immortality, of generating a way of leaving Earth that would perpetuate the light of human consciousness even after this planet is uninhabitable. To some spacefarers and space proponents, the survival of the species is the endgame. Or, put another way, space could be our ultimate salvation.

The return to the Moon should require a reckoning with these ideas. The Moon deserves a conversation about its own legacy, and its own meaning to us, among any discussion of geopolitics. We owe it to the Moon, and to ourselves.

In the early 2020s, private companies are the de facto arbiters of the space surrounding Earth. The rocket company SpaceX controls nearly one-fourth of all space launches, and by 2022 had lofted 3,400 miniature internet-broadcasting satellites into near-Earth orbit. Private companies aim to launch satellites and scientific equipment to the Moon with a regular cadence by the mid-2020s, but existing international governance with respect to the Moon is vague. Less than two years before Apollo 11 lifted off, the United States, United Kingdom, and Soviet Union signed the Outer Space Treaty, which stipulates that no nation can claim territory on a celestial body, that space exploration is for the benefit of all nations, and that exploration must be conducted peacefully. But the treaty contains language that private companies—and the American government—are eyeing as a means to protect their territory nonetheless. The clauses call for the signatories (113 parties and 23 signatories, as of March 2023) to avoid harming one another's science experiments or Moon bases. This means in principle that no one can land on them, for instance, or even land near enough that existing equipment might be damaged.

As Alan Bean and Pete Conrad learned during Apollo 12, landing on the Moon and lifting off from it kicks up a considerable dust storm, and fast-flying lunar dust can scour a spacecraft—or a habitat building—like a Brillo pad. So the Outer Space Treaty's noninterference clauses could essentially mean whoever lands on whatever lunar location first has a claim to it, maybe for good. This is going to be a problem moving forward, because several nations and private companies are interested in the same areas.

Many different countries and private corporations are planning missions in the coming years that will focus on the same regions, often the poles and especially the South Pole–Aitken Basin. The Moon's permanently shadowed craters, areas where the Moon's orbital angle is so extreme and mountains are so high that the Sun never reaches their floors, would be another prime target. So, too, the Peaks of Eternal Light, a few of Thomas Harriot's "promontoryes" where the Sun always shines. In permanently shadowed craters, water ice may remain locked in the dust, where machines could theoretically be installed to extract it and use it for rocket fuel or other purposes. On the Peaks of Eternal Light, solar arrays could soak up sunlight to power rovers or almost any kind of equipment.

Anyone who places equipment on these areas, then, could always have access to it—and no one else would be able to touch it, according to some interpretations of the Outer Space Treaty. Imagine someone erecting a solar telescope on one of the eternally sunny promontories. "Its operation would require non-disturbance, and hence that the Peak remains unvisited by others, effectively establishing a claim," notes Martin Elvis, an astronomer at the Harvard–Smithsonian Center for Astrophysics, in a 2016 research paper.[2]

NASA is also making its intentions in the twenty-first century C.E. clear, and this time, they're not about going for all mankind. In 2020, NASA announced its own new system of bilateral agreements, some of which underscore the Outer Space Treaty and some of which are new. The agency's agreements would require peaceful uses of space, protection of heritage locations like the Apollo landing sites and the final resting places of various rovers, and agreements for things like space debris and space rescues. But the accords contain two lines that anyone who cares about the Moon ought to note. One says that space systems ought to be "interoperable," which means bases could use universal docking systems or other sharable technology. This might favor American companies, accustomed to building sharable technology for American spaceships. The second clause argues that the Outer Space Treaty allows "extraction and use of space resources." Read: mining.

The Moon, our silvery sister and silent eternal companion, could by the end of the 2020s become a mining outpost like Chile's Atacama Desert or the oil fields of Iraq. Private companies took the first steps toward this process in late 2022, with the launch of a privately funded Japanese lander carrying a rover from the United Arab Emirates (though the mission failed in a crash landing). Nations from India to China to the United States plan to send astronauts back by the mid-2020s. In April 2023, NASA announced the first crew that would orbit the Moon since Apollo, and the Artemis III mission, which aims to return humans to the Moon for the first time since 1972, is set to launch sometime in 2025. That means humanity doesn't have much time to decide whether lunar development and even mining are things we all support.

SOME OF THE most enthusiastic lunar exploration advocates hope for a return to the Moon soon, within their lifetimes, to start up a new Moon-based economy. Billionaires like Amazon's Jeff Bezos are planning missions, and government entities like the European Space Agency even hope for an entire Moon Village populated by astronauts, scientists, and entrepreneurs. Water would probably be the most valuable resource to begin with, because its hydrogen and oxygen can be cleaved into the ingredients of rocket fuel—a legacy of Moon fans Konstantin Tsiolkovsky and Hermann Oberth, who invented multistage liquid-fueled rockets. Water prospecting is likely to draw diviners to the Moon's permanently shadowed craters. Under NASA's Commercial Lunar Payload Services program, private firms are competing for grants to design spacecraft that can deliver various landers and instruments, including some that can search for resources like water.

"The economic benefit of Apollo was huge. It put us on a whole new path for miniaturization in our electronics. Our cellphones are because of Apollo," Clive Neal, a lunar geochemist at Notre Dame and a lunar exploration advocate, told me. But government investment on that scale is unlikely to happen again—and moreover, this is still a mostly capitalist society driven by profit. "New commercial involvement is what is going to make it sustainable," Neal said.

One of Moon mining's biggest advocates, Paul Spudis, argued that the Moon's own resources would enable us to become a truly spacefaring civilization. There is "no El Dorado on the Moon," he wrote in his 1996 book, *The Once and Future Moon*. "Nevertheless, the Moon is an object of incredible potential wealth."

The Moon's regolith contains materials like oxygen, helium-3, silicon, aluminum, and iron, all of which could be refined into things like fuel, building materials, and solar panels. As Spudis often put it, profit on the Moon would not necessarily mean money in the bank. Profit could mean the ability to obtain something in space, like rocket fuel, which would be far less costly than carrying everything you need from Earth.

Until he died in 2018, Spudis also argued passionately for a return to the Moon for science. It is a natural laboratory for planetary science, just a three-day flight away. He argued we should use it to study the Sun, the Moon, Earth, and ourselves.

THE VAST MAJORITY of people who study the Moon, and the vast majority of people who are concerned with its caretaking and future, are still scientists like Spudis. Many of them hope the Moon may become some sort of scientific preserve, where people can visit under extreme, austere conditions and conduct research for the benefit of everyone. Under this scenario, the Moon would be much like Antarctica. It would be like an eighth continent shorn from Earth, set apart from the seven continents Wegener's plate tectonics gave us in the present day.

China is already focused on the Moon's far side, developing the hardware it will need to land its taikonauts there. For now, the Chinese Space Agency is focused on research, and its pronouncements are famously coy, but many American observers think China's ambitions include mining and an eventual settlement. In 2018, the country accelerated development of its Long March 9 rocket, similar in size to the Saturn V that launched the Apollo missions. Chinese officials have said the rocket will power its first lunar surface missions in the 2030s. In January 2019, the *Chang'e 4* spacecraft and its *Yutu-2* "Jade Rabbit" rover landed inside the South Pole–Aitken Basin, the first mission to ever land on the Moon's far side. It was a towering achievement, and a major step toward China's long-term Moon plans. The Chinese Space Agency unveiled a new emblem for its lunar exploration program in time for the *Chang'e 4* launch. It is a calligraphic crescent embracing two hash marks, and at a glance it looks like the Chinese character for "Moon," 月. The two hashes are not just marks, but gray bootprints, a clear sign of China's intent.

Plenty of other governments are also eyeing the far side for science.

In 2019, an Italian scientist named Claudio Maccone, of the National Institute for Astrophysics in Italy, called for a radio-free zone on the far side of the Moon. The Moon itself would block the entire hemisphere from any Earthly interference, allowing for operation of the solar system's largest radio telescope, a new observatory that could peer into the darkest heart of the cosmos. In April 2020, NASA funded a study to explore building just such a telescope. The Lunar Crater Radio Telescope would be mounted inside a three-mile-wide crater on the far side of the Moon, and it could look at the universe in ultra-low-frequency wavelengths, especially the types that can't make it through Earth's atmosphere. In 2023, the space agency funded an even bigger telescope, which will use the Moon's own soil to enable on-site construction of solar cells, antenna parts, and power lines. The FarView telescope will be able to see back to the so-called cosmic dark ages, when the first stars ignited—a time that has so far been invisible, even to the powerful James Webb Space Telescope.

The Moon's far side, shielded from Earthly emissions or atmospheric vapors, is a prime location to study the earliest days of the universe. In China, scientists are studying satellite arrays that would fly in formation around the Moon to make similar observations. But even purely exploration-driven Moon bases might invoke some of the Outer Space Treaty's non-interference clauses. Nothing happens in isolation.

THOUGH MUCH OF the lunar exploration agenda of the early 2020s is focused on science, the fundamental questions astronomers seek to answer are just one avenue on the greater road of discovery. Science is only one way of coming to know the universe, but it has the most power, Lisa Messeri, an anthropologist at Yale University, told me. "This becomes a tricky argument. I am someone who believes in the science of climate change, and the science of vaccination. I am also someone who wonders what it means to take a Native American cosmology and treat it equally to our scientific cosmology, when it comes to the Moon."

Who gets to decide the way to use a precious thing, a limited, special, spectral, spiritual thing, that we all share?

THERE IS ANOTHER perspective to consider. That of the Moon itself. It is a place, and it exists nearby but not impassively. "It has cultural and historical agency," as Messeri put it. The Moon does not have feelings, of course. But Spudis himself often spoke of its intrinsic value. A British astronomer named Vera Assis Fernandes has written that the Moon should be acknowledged as an entity deserving of respect.

"The celestial body closest to the Earth is an important, powerful and fragile environment that needs to be understood and taken into consideration before we finally set sail to it again," she wrote in a 2019 research paper. "Have we ever asked why humans want to return to the Moon and then colonise it?"[3]

There is a way forward that considers these questions and answers them in dissent. We don't need to build a human settlement on the Moon. We don't need to do anything at all to the Moon. The Moon cannot speak for itself. We have to speak for it. And no one person, no single culture, can speak for everyone who shares the sky. The Moon belongs to everyone, which means it belongs to no one.

○ ˙

WHEN I WAS writing this book about our relationship with the Moon, I originally planned to visit the lunisolar arrangement at the Cahokia Mounds, located ten miles east of St. Louis, my home for ten years. The mounds are the remnants of the great Mississippian culture, people who lived in the American Midwest a millennium ago. At its height, Ca-

hokia was as sprawling as the city of London in the 1600s. It is sometimes referenced as "the City of the Sun." Cahokia was first inhabited
beginning around 700 C.E., and the Mississippian culture eventually
grew to some twenty thousand residents by 1200 C.E. Around 1100, they
built a wooden circle, now nicknamed "Woodhenge," that aligns with
the solstice sun. But the most striking characteristic is the series of Cahokia Mounds, which served a variety of civic and ritual purposes. The
largest mound connects to an elevated embankment, and it aligns with
the southernmost moonrise of the year.

Then my family made a sudden decision to move across the country,
back to Colorado. Once I was in the West, where I grew up, I planned to
visit Canyon de Chelly in Chinle, Arizona, to view prehistoric rock art
and talk to Navajo and Hopi traditional practitioners. The Hopi people
in the 1960s had celebrated the Apollo Moon landings, because a legend
in their folklore speaks of a new beginning when the Eagle lands on the
Moon. But I had also read that some Navajo people were offended by the
missions, believing that human footprints on the Moon would be akin
to desecration of a holy site. When the father of planetary science, Eugene Shoemaker, died in 1997, his family and NASA officials placed an
ounce of his ashes on board the Lunar Prospector mission. The satellite
was intentionally crashed on the Moon after almost two years in orbit,
consigning some of Shoemaker to the Moon forever. The Navajo Nation
president at the time, Albert Hale, called the act "sacrilegious, a gross
insensitivity to the beliefs of many Native Americans, to place human
remains on the Moon."[4] I wanted to talk to the Dine people, who most
white European descendants call the Navajo, about their vision of the
Moon. And then the coronavirus pandemic shut everything down. Navajo lands were closed to outsiders as the virus rampaged through the
population.

As an alternative, I planned a road trip to the Bighorn Medicine
Wheel, an ancient stone circle in northwestern Wyoming, built between
three hundred and eight hundred years ago by the Plains Indians like
the Cheyenne and Pawnee. It is visible only during the summer, usually
covered in snow through most of the winter months. It contains an arrangement of cairns and spokes that are thought to align with the summer solstice and the heliacal rise of certain stars. And even more

intriguingly, the wheel contains twenty-eight spokes. There are about twenty-eight days in a lunar cycle, and I wanted to explore the ruin and learn the Moon traditions of the Plains Indians who lived in this area before me. But most of the West was under a stay-at-home order during the beginning of the pandemic. It was too far to drive.

It was after the stay-at-home orders were lifted that I brought my family with me to Chimney Rock. We could not go at night, because of the pandemic and because it would have been unsafe for my then five-year-old.

Under the summer Sun, I walked past the Great House foundation and toward the cliff, toward the towers. I wanted to give myself a minute to stare at them, wondering if I would feel something like I felt at Tomnaverie. The ghosts in that lunar monument were quiet. It was not unlike stepping into a very old church, or walking into an uncrowded old-growth forest; the experience was humbling and awesome, in the old sense of that word. I thought if it was quiet enough, I might be faced with something similar at Chimney Rock and forge a connection with the people who built this place. Maybe I could peer through some kind of portal of consciousness to see the Moon they knew, and the sky they watched.

But my daughter was so excited to be walking with me, and so enthusiastic about the small cacti on the trail, and so eager to look for animals hiding in the rocks. She chattered nonstop, as ever. Crackling with kindergartner energy, she ran toward the foundation of the Great House and approached a docent, who was there to answer questions and to warn people where they could and could not step. She climbed down into the house at the docent's invitation, trying to make her voice echo, asking questions about the staircase and the brick walls and why there was no longer a ceiling. There was no such thing as quiet. I realized this was exactly as it should be.

Ten centuries earlier, another young family probably walked through the same house. They brought supplies and snacks like we did, although instead of hydration backpacks, they may have carried woven baskets and painted clay pottery. There were probably lots of kids running around, shouting to their cousins and friends, playing games gathered under the Great House's adobe roof. They might have been there for a celebration, maybe a feast that meant a great deal to them but which we don't understand anymore. I think it's fair to assume the village was not at all quiet.

The people who lived here before me looked different from me and my family. But I feel confident saying we shared certain things, and that some of them have remained the same through the last millennia, or through any span of time. They probably traveled to Chimney Rock with their families. The children were probably restless from a long journey, and wanted a chance to play. They probably brought toys, like my daughter's stash in the back seat of my car. In Ennigaldi's museum at Ur, the diggers found clay dog figurines that were painted red, likely toys for the students at the sacred Giparu. The parents were probably excited to show their children the special places, and to teach what the places meant to their forebears and to the parents themselves. The parents probably worried about their children approaching the cliff's edge. A mother like me might have smiled at her firstborn's precociousness. She may have stood beside the Great House, watching her oldest, thinking about her new child yet to be born, and wondering if she would be fierce and independent like her sister. My experience mirrored their experience, which mirrored even older experiences, which have always repeated themselves, under the same Sun, under the same Moon.

JUST AS IT was for the Puebloans, just as it was for Enheduanna and Nabonidus, just as it was for the Greek philosophers and Johannes Kepler and Wernher von Braun and Michael Collins, the Moon can still be special and untouchable. It can remain a spectral reminder in the night. It can still be an omen worth celebrating. It can be joyous to witness, full and glittering on a freezing winter night, or a surprising crescent peeking through summer-leafed trees. It can sneak up on us; most Moons you will see throughout your life are gibbous, not quite there but not quite gone. The Moon is only full and bright for two or three days a month. If you're lucky, you will see a few hundred of these in your life. Once in a while you might catch it hanging overhead as a scimitar at dusk. It might startle you, yellowish and half dark on the eastern horizon, if you awaken in the hours before dawn. It might sail ahead of the Sun hours before your alarm clock rings, and hang low in the west in the morning as you leave for work or school. It might fade into the background, like white noise, like the stars you may be lucky enough to see at night.

But wherever you are, whatever you believe, from whichever, if any, descendant of Abraham of Ur you take your tradition, you will always be able to trust in the Moon. It will always be above you. It will always return, quietly lighting the night for you, peering through the clouds in the morning for you. You will be able to count on its inconstant appearance. You can be sure that it will rise, cool and glimmering, sometime tomorrow. You'll be able to see it from wherever you are here on Earth, the Moon's home, its eternal companion, its cloud-wisped sister and reason for being. Walk outside under the Moon tonight or tomorrow morning. Look up, walk along with it, and say hello.

ACKNOWLEDGMENTS

MANY WRITERS LIKE to complain about the lonesome toil of creating their art. I do not agree with them. A nonfiction book may be constructed in a writer's head, often in the middle of the night, but this work would be impossible without friends, colleagues, and readers. Writing a book is one of the most communal tasks I can imagine, and I am so thankful to you, reader, especially if you've gotten this far.

I am eternally grateful to my wonderful agent, Laurie Abkemeier, for her insight and boundless patience. A hundred thousand thanks to my brilliant editor, Hilary Redmon, a generous, intellectually curious writer's editor I am so lucky to work with, and who saw what I wanted to do with this book better than I did. I am also grateful to my editors at Sceptre, Juliet Brooke and Jo Dingley, who offered helpful insight and narrative advice.

The first seeds of this book germinated in the Peakview Elementary School library, where I sat on the floor as a student and listened to the Apollo 11 recordings. I am indebted to my history teachers, especially Doug Chilton, who first taught me about the Sumerians; Deborah Milliser, who gave me Howard Zinn when I was sixteen; and Elizabeth Jones, who showed me new perspectives on Western history. I am also grateful to Susan Wise Bauer for reminding me that history writing can be stylish and fun as well as comprehensive. And I am grateful to my parents, for encouraging me in all things but especially for fostering my love of books and of space for as long as I can remember. The Halley's Comet 1986 T-shirt I wore in kindergarten is probably in a basement storage box somewhere, my official U.S. Space Camp windbreaker is still in my closet, and my autographed photo of Neil Armstrong, courtesy of my dad, now hangs in my home office.

Many scientists, academics, museum curators, assistants, archivists, and librarians helped me piece together this narrative and guided me on my travels. I am especially grateful to Hilary and Charlie Murray, for opening their home and work to me; to Harald Meller, Konstanze Geppert, Georg Schafferer, and Bettina Pfaff of the State Museum of Prehistory in Halle, Germany; Alison McCann, archivist at Petworth House, and Lord Egremont; Don Olson; Vince Gaffney; Willis Monroe; Hannah Byrne; Per Ahlberg; Kristin Tessmar-Raible; Charlotte Helfrich-Förster; Anthony Aveni; Andrea Mosie; Ryan Ziegler; Charis Krysher; Andrea Jones; Teasel Muir-Harmony; Roger Launius; Brad Joliff; Clive Neal; Bill Bottke; Sarah T. Stewart; Simon Lock; Dave Stevenson; Dean Swinford; J. McKim Malville; and Daniel Graham. Many others also helped me at Johnson Space Center, the British Museum, the galleries and museums of Berlin's Museum Island, Chimney Rock National Monument, the Charles L. Tutt Library at Colorado College, the Pikes Peak State College Library, the Pikes Peak Library District, the John M. Olin Library at Washington University in St. Louis, and the National Archives' National Personnel Records Center in St. Louis. I'm grateful to my fact-checker, Maya Dusenbery; and Kaeli Subberwal, Miriam Khanukaev, and the editorial teams at Random House in the United States and Sceptre in the United Kingdom, especially the extremely thorough copy desk and production teams. Any errors that remain are mine. If I left anyone out, I apologize; feel free to be newly annoyed at me every time the Moon is visible.

Thanks to Tom Kearney, Mary Allen, Chris Cobler, and Randy Bangert for giving me a start, and to every editor who accepted a story after I left daily newspapers, especially Paul Adams, Susannah Locke, Brigid Hains, Ross Andersen, Michelle Nijhuis, Michael Moyer, Alan Burdick, Michael Roston, Lee Billings, and Clara Moskowitz.

Large portions of this book were written and revised at night while my children were sleeping. My deep appreciation goes to Lin-Manuel Miranda, Elton John, Kristen Anderson-Lopez, Robert Lopez, and Chris Ballew aka Caspar Babypants for providing the soundtrack to my life. I am indebted to Kate Werner for watching my older daughter during the bizarre summer of 2020, and then caring for my newborn the following year.

I work alone in my home office, an arrangement I would never survive without my dog, Sunshine, or my many supportive colleagues, especially Christie Aschwanden; Lisa Grossman; Sarah Scoles; Julia Rosen; Alex Witze; Virginia Hughes and the Gentlewomen; Jeanne Erdmann, for introducing me to so many of these fine people; Swapna Krishna and everyone in the Miss Guided Missiles; Adam Rogers, Maryn McKenna, Emily Willingham, and the other talented journalists of The Thing; Jersey Knit Goals; the writers of the group blog The Last Word on Nothing; Andrew Curry; David Brown; Ferris Jabr; Ed Yong; and so many others. I am most grateful to Katharine Gammon, for wisdom, wit, and camaraderie while navigating motherhood and a journalism career, and to Peter Brannen, for introducing me to my agent, for sending research materials, and for never being more than a chat window away over the years. *Sláinte.*

Finally, this book absolutely would not have happened without the support of my family, especially my parents, Nancy and Doug; my husband, Greg; and my two daughters. Abigail and now Evelyn have both loved the Moon since their earliest times on this planet, and I pray it will give them comfort throughout all of their days.

May the Moon's appearance in the sky do the same for you.

"Across the Reef: The Assault on Betio." *Naval History Magazine* 22, no. 6 (December 2008). Accessed March 14, 2022. https://www.usni.org/magazines/naval-history-magazine/2008/december/across-reef-assault-betio.

Aldersey-Williams, Hugh. *The Tide: The Science and Stories Behind the Greatest Force on Earth.* New York: W. W. Norton, 2017.

Aldrin, Buzz, and Ken Abraham. *Magnificent Desolation: The Long Journey Home from the Moon.* Reprint ed. New York: Three Rivers Press, 2010.

Alexander, Joseph. *Across the Reef: The Marine Assault of Tarawa.* Washington, D.C.: U.S. Marine Corps Headquarters, History and Museums Division, 2013.

Allen, Reginald E., ed. *Greek Philosophy: Thales to Aristotle.* Rev. ed. New York: Free Press, 1991.

Ambrose, Stephen E. *D-Day: June 6, 1944: The Climactic Battle of World War II.* Reprint ed. New York: Simon & Schuster, 2013.

———. *Pegasus Bridge.* New York: Touchstone Books, 1988.

Appian. *The Civil Wars.* Translated by John Carter. Penguin Classics. New York: Penguin Books, 1996.

Arianrhod, Robyn. *Thomas Harriot: A Life in Science.* New York: Oxford University Press, 2019.

Armstrong, John C., Llyd E. Wells, and Guillermo Gonzalez. "Rummaging Through Earth's Attic for Remains of Ancient Life." *Icarus* 160, no. 1 (November 1, 2002): 183–96. https://doi.org/10.1006/ICAR.2002.6957.

Asimov, Isaac. *The Tragedy of the Moon.* New York: Dell, 1978.

Asphaug, Erik. *When the Earth Had Two Moons: Cannibal Planets, Icy Giants, Dirty Comets, Dreadful Orbits, and the Origins of the Night Sky.* Reprint ed. New York: Custom House, 2019.

Aveni, Anthony. *The Book of the Year: A Brief History of Our Seasonal Holidays.* New York: Oxford University Press, 2002.

———. *Empires of Time: Calendars, Clocks, and Cultures.* New York: Basic Books, 1989.

———. *People and the Sky: Our Ancestors and the Cosmos.* London: Thames & Hudson, 2008.

Barboni, Melanie, et al. "Early Formation of the Moon 4.51 Billion Years Ago." *Science Advances* 3, no. 1 (January 11, 2017): e1602365. https://doi.org/10.1126/sciadv.1602365.

Barbree, Jay, and John Glenn. *Neil Armstrong: A Life of Flight.* New York: Thomas Dunne Books, 2014.

Bates, Bryan C. "Hopi and Anasazi Alignments and Rock Art." In *Handbook of Archaeoastronomy and Ethnoastronomy,* ed. C. L. N. Ruggles, 607–19. New York: Springer, 2015. https://doi.org/10.1007/978-1-4614-6141-8_42.

Battaglia, Pietro, et al. "Influence of Lunar Phases, Winds and Seasonality on the Stranding of Mesopelagic Fish in the Strait of Messina (Central Mediterranean Sea)." *Marine Ecology* 38, no. 5 (October 1, 2017). https://doi.org/10.1111/MAEC.12459.

Bauer, Susan Wise. *The History of the Ancient World: From the Earliest Accounts to the Fall of Rome.* New York: W. W. Norton, 2007.

———. *The History of the Medieval World: From the Conversion of Constantine to the First Crusade.* New York: W. W. Norton, 2010.

———. *The Story of Western Science: From the Writings of Aristotle to the Big Bang Theory.* New York: W. W. Norton, 2015.

Beard, Mary. *SPQR: A History of Ancient Rome.* Reprint ed. New York: Liveright, 2016.

Beaulieu, Paul-Alain. *A History of Babylon, 2200 BC–AD 75.* Hoboken, N.J.: Wiley-Blackwell, 2018.

———. *Reign of Nabonidus, King of Babylon (556–539 B.C.).* New Haven, Conn.: Yale University Press, 1989. https://doi.org/10.2307/j.ctt2250wnt.

Bedrosian, T. A., and R. J. Nelson. "Influence of the Modern Light Environment on Mood." *Molecular Psychiatry* 18, no. 7 (July 2013): 751–57. https://doi.org/10.1038/mp.2013.70.

Belmonte, Juan Antonio. "Ancient 'Observatories'—A Relevant Concept?" In *Handbook of Archaeoastronomy and Ethnoastronomy,* ed. C. L. N. Ruggles, 133–45. New York: Springer, 2015. https://doi.org/10.1007/978-1-4614-6141-8_9.

———. "Solar Alignments—Identification and Analysis." In *Handbook of Archaeoastronomy and Ethnoastronomy,* ed. C. L. N. Ruggles, 483–92. New York: Springer, 2015. https://doi.org/10.1007/978-1-4614-6141-8_36.

Benson, L. V., E. M. Hattori, J. Southon, and B. Aleck. "Dating North America's Oldest Petroglyphs, Winnemucca Lake Subbasin, Nevada." *Journal of Ar-*

chaeological Science 40, no. 12 (December 1, 2013): 4466–76. https://doi.org/10.1016/J.JAS.2013.06.022.

Bergaust, Erik. *Wernher von Braun: The Authoritative and Definitive Biographical Profile of the Father of Modern Space Flight.* Washington, D.C.: National Space Institute, 1976.

Bertman, Stephen. *Handbook to Life in Ancient Mesopotamia.* New York: Oxford University Press, 2005.

Biddle, Wayne. *Dark Side of the Moon: Wernher von Braun, the Third Reich, and the Space Race.* New York: W. W. Norton, 2009.

Bloom, Terrie F. "Borrowed Perceptions: Harriot's Maps of the Moon." *Journal for the History of Astronomy* 9, no. 2 (June 22, 1978): 117–22. https://doi.org/10.1177/002182867800900203.

Borman, Frank, and Robert J. Serling. *Countdown: An Autobiography.* Norwalk, Conn.: Easton Press, 1997.

Bostwick, Todd W. "Hohokam Archaeoastronomy." In *Handbook of Archaeoastronomy and Ethnoastronomy,* ed. C. L. N. Ruggles, 551–64. New York: Springer, 2015. https://doi.org/10.1007/978-1-4614-6141-8_43.

Boyle, Rebecca. "These Dusty Young Stars Are Changing the Rules of Planet-Building." *Nature* 564, no. 7734 (December 4, 2018): 20–23. https://doi.org/10.1038/d41586-018-07591-8.

Bradley, Richard. *A Geography of Offerings: Deposits of Valuables in the Landscapes of Ancient Europe.* Philadelphia: Oxbow Books, 2017.

———. *The Moon and the Bonfire: An Investigation of Three Stone Circles in North-East Scotland.* Edinburgh: Society of Antiquaries of Scotland, 2005.

———. *The Significance of Monuments: On the Shaping of Human Experience in Neolithic and Bronze Age Europe.* New York: Routledge, 1998.

Braun, Wernher von, and Thomas O. Paine. *The Mars Project.* Reprint ed. Urbana: University of Illinois Press, 1962.

Caesar, Julius. *The Civil War of Caesar.* Translated by Jane P. Gardner. Reprint ed. Penguin Classics. Harmondsworth, U.K.: Penguin Books, 1976.

———. *The Gallic Wars.* Translated by Thomas Holmes. N.p.: CreateSpace, 2016.

———. *Seven Commentaries on the Gallic War with an Eighth Commentary by Aulus Hirtius.* Oxford: Oxford University Press, 1996.

Campion, Nicholas. "Astrology as Cultural Astronomy." In *Handbook of Archaeoastronomy and Ethnoastronomy,* ed. C. L. N. Ruggles, 103–16. New York: Springer, 2015. https://doi.org/10.1007/978-1-4614-6141-8_16.

———. *Astrology and Popular Religion in the Modern West: Prophecy, Cosmology and the New Age Movement.* New York: Routledge, 2016.

Caspar, Max. *Kepler.* New York: Dover, 2012.

Cerveny, Randall S., and Robert C. Balling. "Lunar Influence on Diurnal Temperature Range." *Geophysical Research Letters* 26, no. 11 (June 1, 1999): 1605–7. https://doi.org/10.1029/1999GL900303.

Chamberlain, Von Del. "Diné (Navajo) Ethno- and Archaeoastronomy." In *Handbook of Archaeoastronomy and Ethnoastronomy*, ed. C. L. N. Ruggles, 629–40. New York: Springer, 2015. https://doi.org/10.1007/978-1-4614-6141-8_44.

Chapman, Allan. "A New Perceived Reality: Thomas Harriot's Moon Maps." *Astronomy and Geophysics* 50, no. 1 (February 1, 2009): 1.27–1.33. https://doi.org/10.1111/j.1468-4004.2009.50127.x.

Churchill, Winston, *The Second World War*, vol. 5: *Closing the Ring*. Boston: Houghton Mifflin, 1951.

Cline, Eric H. *1177 B.C.: The Year Civilization Collapsed*. Rev. ed. Princeton, N.J.: Princeton University Press, 2021.

Collins, Michael. *Carrying the Fire: An Astronaut's Journeys*. 50th anniversary ed. New York: Farrar, Straus & Giroux, 2019.

Comins, Neil F. *What If the Moon Didn't Exist?: Voyages to Earths That Might Have Been*. New York: HarperCollins, 1993.

Connor, James A. *Kepler's Witch: An Astronomer's Discovery of Cosmic Order amid Religious War, Political Intrigue, and the Heresy Trial of His Mother*. New York: HarperCollins, 2009.

Cotte, Michel. "Archaeoastronomical Heritage and the World Heritage Convention." In *Handbook of Archaeoastronomy and Ethnoastronomy*, ed. C. L. N. Ruggles, 301–11. New York: Springer, 2015. https://doi.org/10.1007/978-1-4614-6141-8_18.

Crawford, Harriet. *Sumer and the Sumerians*. 2nd ed. New York: Cambridge University Press, 2004.

Cullen, Christopher. "The First Complete Chinese Theory of the Moon: The Innovations of Liu Hong c. A.D. 200." *Journal for the History of Astronomy* 33 (February 1, 2002): 21–39. https://doi.org/10.1177/002182860203300104.

———. *Heavenly Numbers: Astronomy and Authority in Early Imperial China*. Oxford: Oxford University Press, 2017.

Dalley, Stephanie, trans. *Myths from Mesopotamia: Creation, the Flood, Gilgamesh, and Others*. Rev. ed. Oxford: Oxford University Press, 2009.

Daly, Reginald A. "Origin of the Moon and Its Topography." *Proceedings of the American Philosophical Society* 90, no. 2 (1946): 104–19. https://www.jstor.org/stable/3301051.

Darwin, Charles. *The Origin of Species*. 150th anniversary ed. New York: Signet, 2009.

Darwin, George Howard. *The Tides and Kindred Phenomena in the Solar System*. Boston: Houghton Mifflin, 1899.

Della Monica, Ciro, Giuseppe Atzori, and Derk Jan Dijk. "Effects of Lunar Phase on Sleep in Men and Women in Surrey." *Journal of Sleep Research* 24, no. 6 (December 1, 2015): 687–94. https://doi.org/10.1111/jsr.12312.

Dijk, Derk Jan, and Anne C. Skeldon. "Biological Rhythms: Human Sleep Before the Industrial Era." *Nature* 527, no. 7577 (November 11, 2015): 176–77. https://doi.org/10.1038/527176A.

Eisenhower, Dwight D. *Crusade in Europe: A Personal Account of World War II.* New York: Doubleday, 1948.

Elardo, Stephen M., Matthieu Laneuville, Francis M. McCubbin, and Charles K. Shearer. "Early Crust Building Enhanced on the Moon's Nearside by Mantle Melting-Point Depression." *Nature Geoscience* 13, no. 5 (May 2020): 339–43. https://doi.org/10.1038/s41561-020-0559-4.

Eliade, Mircea. *The Sacred and the Profane: The Nature of Religion.* Translated by Willard R. Trask. San Diego: Harcourt Brace Jovanovich, 1987.

Emo, Robert B., et al. "Evidence for Evolved Hadean Crust from Sr Isotopes in Apatite Within Eoarchean Zircon from the Acasta Gneiss Complex." *Geochimica et Cosmochimica Acta* 235 (August 15, 2018): 450–62. https://doi.org/10.1016/J.GCA.2018.05.028.

Fagan, Brian. *From Black Land to Fifth Sun: The Science of Sacred Sites.* Reading, Mass.: Basic Books, 1999.

———. *Return to Babylon: Travelers, Archaeologists, and Monuments in Mesopotamia.* Boulder: University Press of Colorado, 2007.

Feng, Li. *Early China: A Social and Cultural History.* New York: Cambridge University Press, 2013.

Fisher, Victor B. "Presentation of Archaeoastronomy in Introductions to Archaeology." In *Handbook of Archaeoastronomy and Ethnoastronomy,* ed. C. L. N. Ruggles, 251–61. New York: Springer, 2015. https://doi.org/10.1007/978-1-4614-6141-8_12.

Fitzpatrick, Andrew P., and Colin Haselgrove, eds. *Julius Caesar's Battle for Gaul: New Archaeological Perspectives.* Oxford: Oxbow Books, 2019.

Fox, Robert, ed. *Thomas Harriot and His World.* New York: Routledge, 2016.

Frank, Roslyn M. "Origins of the 'Western' Constellations." In *Handbook of Archaeoastronomy and Ethnoastronomy,* ed. C. L. N. Ruggles, 147–63. New York: Springer, 2015. https://doi.org/10.1007/978-1-4614-6141-8_11.

Galilei, Galileo. *Dialogue Concerning the Two Chief World Systems: Ptolemaic and Copernican.* Translated by Stillman Drake. New York: Modern Library, 2001.

———. *Sidereus nuncius.* N.p.: Apud Thomam Baglionum, 1610. https://doi.org/10.5479/sil.95438.39088015628597.

Garrick-Bethell, Ian, et al. "Troctolite 76535: A Sample of the Moon's South Pole–Aitken Basin?" *Icarus* 338 (March 1, 2020): 113430. https://doi.org /10.1016/j.icarus.2019.113430.

Gates, Charles. *Ancient Cities: The Archaeology of Urban Life in the Ancient Near East and Egypt, Greece and Rome.* New York: Routledge, 2003.

Gavin, James Maurice. *Aerial Warfare.* Washington, D.C.: Infantry Journal Press, 1947.

———. *Airborne Warfare.* Lulu.com, 2020.

Gee, Henry. "Moonlight and Global Warming." *Nature News,* June 24, 1999. https://doi.org/10.1038/news990624-9.

Goldin, Paul R., ed. *Routledge Handbook of Early Chinese History.* New York: Routledge, 2020.

González-García, A. César. "Lunar Alignments—Identification and Analysis." In *Handbook of Archaeoastronomy and Ethnoastronomy,* ed. C. L. N. Ruggles, 493–506. New York: Springer, 2015. https://doi.org/10.1007/978-1-4614 -6141-8_37.

Graham, Daniel W. *Explaining the Cosmos: The Ionian Tradition of Scientific Philosophy.* Princeton, N.J.: Princeton University Press, 2006.

———. *Science Before Socrates: Parmenides, Anaxagoras, and the New Astronomy.* Oxford; New York: Oxford University Press, 2013.

Greenbaum, Dorian Gieseler. "Astronomy, Astrology, and Medicine." In *Handbook of Archaeoastronomy and Ethnoastronomy,* ed. C. L. N. Ruggles, 117–32. New York: Springer, 2015. https://doi.org/10.1007/978-1-4614-6141-8_19.

Guthke, Karl. *The Last Frontier: Imagining Other Worlds, from the Copernican Revolution to Modern Science Fiction.* Translated by Helen Atkins. Ithaca, N.Y.: Cornell University Press, 1990, 1993.

Haba-Rubio, José, et al. "Bad Sleep? Don't Blame the Moon! A Population-Based Study." *Sleep Medicine* 16, no. 11 (2015): 1321–26. https://doi.org/10.1016 /j.sleep.2015.08.002.

Hansen, James R. *First Man: The Life of Neil A. Armstrong.* Reissue ed. New York: Simon & Schuster, 2018.

Hartmann, William K., and Donald R. Davis. "Satellite-Sized Planetesimals and Lunar Origin." *Icarus* 24, no. 4 (April 1975): 504–15. https://doi.org /10.1016/0019-1035(75)90070-6.

Henry, John. *Moving Heaven and Earth (Icon Science): Copernicus and the Solar System.* London: Icon Books, 2017.

Herman, Arthur. *The Cave and the Light: Plato Versus Aristotle, and the Struggle for the Soul of Western Civilization.* New York: Random House, 2013.

Hetherington, Norriss S., ed. Encyclopedia of Cosmology: Historical, Philo-

sophical, and Scientific Foundations of Modern Cosmology. New York: Routledge, 2015.

Holmes, Richard. *The Age of Wonder: How the Romantic Generation Discovered
the Beauty and Terror of Science.* New York: Vintage, 2009.

Hoyt, Edwin P. *Storm over the Gilberts: War in the Central Pacific, 1943.* New York:
Avon Books, 1983.

Hurt, Harry, III. *For All Mankind.* Reprint ed. New York: Grove Press, 2019.

Ilias, I., et al. "Do Lunar Phases Influence Menstruation? A Year-Long Retro-
spective Study." *Endocrine Regulations* 47, no. 3 (2013): 121–22. https://doi
.org/10.4149/endo_2013_03_121.

Irving, Washington. *A History of New York.* 1809; reprint London: Penguin Books,
2008.

———. *Knickerbocker's History of New York, Complete.* Chicago: W. B. Conkey,
1809. https://www.gutenberg.org/files/13042/13042-h/13042-h.htm.

Irwin, James B. *More Than Earthlings: An Astronaut's Thoughts for Christ-
Centered Living.* Nashville, Tenn.: Broadman Press, 1983.

Iwaniszewski, Stanisław. "Astrotourism and Archaeoastronomy." In *Handbook
of Archaeoastronomy and Ethnoastronomy,* ed. C. L. N. Ruggles, 287–300. New
York: Springer, 2015. https://doi.org/10.1007/978-1-4614-6141-8_21.

———. "Concepts of Space, Time, and the Cosmos." In *Handbook of Archaeoastronomy and Ethnoastronomy,* ed. C. L. N. Ruggles, 3–14. New York: Springer,
2015. https://doi.org/10.1007/978-1-4614-6141-8_2.

———. "Cultural Interpretation of Archaeological Evidence Relating to Astronomy." In *Handbook of Archaeoastronomy and Ethnoastronomy,* ed. C. L. N.
Ruggles, 315–24. New York: Springer, 2015. https://doi.org/10.1007/978-1
-4614-6141-8_24.

Jacobson, Seth A., et al. "Formation, Stratification, and Mixing of the Cores of
Earth and Venus." *Earth and Planetary Science Letters* 474 (2017): 375–86.
https://doi.org/10.1016/j.epsl.2017.06.023.

Kant, Immanuel. *Universal Natural History and Theory of the Heavens.* Translated by Ian Johnston. Arlington, Va.: Richer Resources Publications, 2009.

Kempenaers, Bart, et al. "Artificial Night Lighting Affects Dawn Song, Extra-Pair
Siring Success, and Lay Date in Songbirds." *Current Biology* 20, no. 19
(October 12, 2010): 1735–39. https://doi.org/10.1016/J.CUB.2010.08.028.

Kepler, Johannes. *Conversation with Galileo's Sidereal Messenger.* Translated and
edited by Edward Rosen. New York: Johnson, 1965.

Kershaw, Alex. *The First Wave: The D-Day Warriors Who Led the Way to Victory in
World War II.* New York: Dutton Caliber, 2019.

King, David A. "Astronomy in the Service of Islam." In *Handbook of Archaeoas-*

tronomy and Ethnoastronomy, ed. C. L. N. Ruggles, 181–96. New York: Springer, 2015. https://doi.org/10.1007/978-1-4614-6141-8_13.

Koestler, Arthur. *The Sleepwalkers: A History of Man's Changing Vision of the Universe.* 1959; reprint London: Hutchinson, 1968.

———. *Watershed: A Biography of Johannes Kepler.* New York: Doubleday Anchor, 1960.

Krauss, Rolf. "Astronomy and Chronology—Babylonia, Assyria, and Egypt." In *Handbook of Archaeoastronomy and Ethnoastronomy,* ed. C. L. N. Ruggles, 31–41. New York: Springer, 2015. https://doi.org/10.1007/978-1-4614-6141-8_3.

Kriwaczek, Paul. *Babylon: Mesopotamia and the Birth of Civilization.* New York: St. Martin's Griffin, 2012.

Krupp, Edwin C. "Archaeoastronomical Concepts in Popular Culture." In *Handbook of Archaeoastronomy and Ethnoastronomy,* ed. C. L. N. Ruggles, 263–85. New York: Springer, 2015. https://doi.org/10.1007/978-1-4614-6141-8_20.

———. "Astronomy and Power." In *Handbook of Archaeoastronomy and Ethnoastronomy,* ed. C. L. N. Ruggles, 67–91. New York: Springer, 2015. https://doi.org/10.1007/978-1-4614-6141-8_5.

———. *Beyond the Blue Horizon: Myths and Legends of the Sun, Moon, Stars, and Planets.* New York: HarperCollins, 1991.

———. *Echoes of the Ancient Skies: The Astronomy of Lost Civilizations.* Dover Books on Astro ed. Mineola, N.Y.: Dover, 2003.

———. "Rock Art of the Greater Southwest." In *Handbook of Archaeoastronomy and Ethnoastronomy,* ed. C. L. N. Ruggles, 593–606. New York: Springer, 2015. https://doi.org/10.1007/978-1-4614-6141-8_45.

Last, Kim S., et al. "Moonlight Drives Ocean-Scale Mass Vertical Migration of Zooplankton During the Arctic Winter." *Current Biology* 26, no. 2 (January 25, 2016): 244–51. https://doi.org/10.1016/j.cub.2015.11.038.

Launius, Roger D. *Apollo's Legacy: Perspectives on the Moon Landings.* Washington, D.C.: Smithsonian Books, 2019.

Lear, John. *Kepler's Dream. With the Full Text and Notes of "Somnium, Sive Astronomia Lunaris," Joannis Kepleri.* Translated by Patricia Frueh Kirkwood. Berkeley: University of California Press, 1965.

Leick, Gwendolyn. *Mesopotamia: The Invention of the City.* London: Penguin Books, 2003.

López, Alejandro Martín. "Cultural Interpretation of Ethnographic Evidence Relating to Astronomy." In *Handbook of Archaeoastronomy and Ethnoastronomy,* ed. C. L. N. Ruggles, 341–52. New York: Springer, 2015. https://doi.org/10.1007/978-1-4614-6141-8_14.

———. "Interactions Between 'Indigenous' and 'Colonial' Astronomies: Adaptation of Indigenous Astronomies in the Modern World." In *Handbook of Ar-*

chaeoastronomy and Ethnoastronomy, ed. C. L. N. Ruggles, 197–212. New York: Springer, 2015. https://doi.org/10.1007/978-1-4614-6141-8_30.

Love, David K. *Kepler and the Universe: How One Man Revolutionized Astronomy.* Essex, Connecticut: Prometheus, 2015.

MacDonald, John. "Inuit Astronomy." In *Handbook of Archaeoastronomy and Ethnoastronomy,* ed. C. L. N. Ruggles, 533–39. New York: Springer, 2015. https://doi.org/10.1007/978-1-4614-6141-8_40.

MacPherson, Hector, Jr. "Kant as an Astronomical Thinker." *Popular Astronomy* 21 (August 1, 1913): 424–27. https://ui.adsabs.harvard.edu/abs/1913PA..... 21..424M/abstract.

Maher, Neil M. *Apollo in the Age of Aquarius.* Cambridge, Mass.: Harvard University Press, 2017.

Mailer, Norman. *Of a Fire on the Moon.* Reprint ed. New York: Random House, 2014.

Makishima, Akio. *Origins of the Earth, Moon, and Life: An Interdisciplinary Approach.* Elsevier, 2017.

Malville, J. McKim, and Andrew Munro. "Great Houses and the Sun—Astronomy of Chaco Canyon." In *Handbook of Archaeoastronomy and Ethnoastronomy,* ed. C. L. N. Ruggles, 577–91. New York: Springer, 2015. https://doi.org /10.1007/978-1-4614-6141-8_47.

Marshack, Alexander. *The Roots of Civilization: The Cognitive Beginnings of Man's First Art, Symbol and Notation.* Rev. ed. Mount Kisco, N.Y.: Moyer Bell, 1991.

McCluskey, Stephen C. "Analyzing Light-and-Shadow Interactions." In *Handbook of Archaeoastronomy and Ethnoastronomy,* ed. C. L. N. Ruggles, 427–44. New York: Springer, 2015. https://doi.org/10.1007/978-1-4614-6141-8_27.

———. "Astronomy in the Service of Christianity." In *Handbook of Archaeoastronomy and Ethnoastronomy,* ed. C. L. N. Ruggles, 165–79. New York: Springer, 2015. https://doi.org/10.1007/978-1-4614-6141-8_15.

———. "Cultural Interpretation of Historical Evidence Relating to Astronomy." In *Handbook of Archaeoastronomy and Ethnoastronomy,* ed. C. L. N. Ruggles, 325–39. New York: Springer, 2015. https://doi.org/10.1007/978-1-4614-6141 -8_29.

———. "Disciplinary Perspectives on Archaeoastronomy." In *Handbook of Archaeoastronomy and Ethnoastronomy,* ed. C. L. N. Ruggles, 227–37. New York: Springer, 2015. https://doi.org/10.1007/978-1-4614-6141-8_23.

———. "Hopi and Puebloan Ethnoastronomy and Ethnoscience." In *Handbook of Archaeoastronomy and Ethnoastronomy,* ed. C. L. N. Ruggles, 649–58. New York: Springer, 2015. https://doi.org/10.1007/978-1-4614-6141-8_48.

McDougall, Walter . . . . *the Heavens and the Earth: A Political History of the Space Age.* New York: Basic Books, 1985.

Meador, Betty De Shong, and Judy Grahn. *Inanna, Lady of Largest Heart: Poems of the Sumerian High Priestess.* Austin: University of Texas Press, 2001.

Meador, Betty De Shong, and John Maier. *Princess, Priestess, Poet: The Sumerian Temple Hymns of Enheduanna.* Austin: University of Texas Press, 2009.

Messeri, Lisa. *Placing Outer Space: An Earthly Ethnography of Other Worlds.* Durham, N.C.: Duke University Press, 2016.

Milosavljevic, Nina. "How Does Light Regulate Mood and Behavioral State?" *Clocks and Sleep* 1, no. 3 (July 12, 2019): 319–31. https://doi.org/10.3390/CLOCKSSLEEP1030027.

Moller, Violet. *The Map of Knowledge: How Classical Ideas Were Lost and Found; A History in Seven Cities.* London: Picador, 2020.

Montgomery, Scott L. *The Moon and the Western Imagination.* Tucson: University of Arizona Press, 1999.

Muir-Harmony, Teasel. *Operation Moonglow: A Political History of Project Apollo.* New York: Basic Books, 2020.

Munson, Gregory E. "Mesa Verde Archaeoastronomy." In *Handbook of Archaeoastronomy and Ethnoastronomy,* ed. C. L. N. Ruggles, 565–75. New York: Springer, 2015. https://doi.org/10.1007/978-1-4614-6141-8_49.

Murray, Hilary K., J. C. Murray, and Caroline Fraser. *A Tale of the Unknown Unknowns: A Mesolithic Pit Alignment and a Neolithic Timber Hall at Warren Field, Crathes, Aberdeenshire.* Oxford: Oxbow Books, 2009.

Murray, William Breen. "Astronomy and Rock Art Studies." In *Handbook of Archaeoastronomy and Ethnoastronomy,* ed. C. L. N. Ruggles, 239–49. New York: Springer, 2015. https://doi.org/10.1007/978-1-4614-6141-8_10.

Nabonidus. *The Abu Habba Cylinder of Nabuna'id, v. Rawlinson pl. 64.* Classic Reprint, 2009.

Nakajima, Miki, and David J. Stevenson. "Inefficient Volatile Loss from the Moon-Forming Disk: Reconciling the Giant Impact Hypothesis and a Wet Moon." *Earth and Planetary Science Letters* 487 (April 1, 2018): 117–26. https://doi.org/10.1016/J.EPSL.2018.01.026.

Naylor, Ernest. *Moonstruck: How Lunar Cycles Affect Life.* Reprint ed. New York: Oxford University Press, 2018.

Neufeld, Michael. *Von Braun: Dreamer of Space, Engineer of War.* Reprint ed. New York: Vintage, 2008.

Neugebauer, Otto. *The Exact Sciences in Antiquity.* 2nd ed. Providence, R.I.: Brown University Press, 1957.

Newitz, Annalee. *Four Lost Cities: A Secret History of the Urban Age.* New York: W. W. Norton, 2021.

Oates, Joan. *Babylon.* London: Thames & Hudson, 1986.

Olson, Don. *Celestial Sleuth: Using Astronomy to Solve Mysteries in Art, History and Literature.* New York: Springer Praxis Books, 2014.

Ordway, Frederick I., III, and Randy Liebermann, eds. *Blueprint for Space: Science Fiction to Science Fact.* Washington, D.C.: Smithsonian Institution Press, 1992.

Packer, Craig, Alexandra Swanson, Dennis Ikanda, and Hadas Kushnir. "Fear of Darkness, the Full Moon and the Nocturnal Ecology of African Lions." *PLoS One* 6, no. 7 (2011): e22285. https://doi.org/10.1371/JOURNAL.PONE.0022285.

Palmer, M. S., et al. "A 'Dynamic' Landscape of Fear: Prey Responses to Spatiotemporal Variations in Predation Risk Across the Lunar Cycle." *Ecology Letters* 20, no. 11 (November 1, 2017): 1364–73. https://doi.org/10.1111/ele.12832.

Pankenier, David W. *Astrology and Cosmology in Early China: Conforming Earth to Heaven.* New ed. Cambridge: Cambridge University Press, 2015.

———. "The Cosmo-Political Background of Heaven's Mandate." *Early China* 20 (1995): 121–76. https://www.jstor.org/stable/23351765.

Pimenta, Fernando. "Astronomy and Navigation." In *Handbook of Archaeoastronomy and Ethnoastronomy,* ed. C. L. N. Ruggles, 43–65. New York: Springer, 2015. https://doi.org/10.1007/978-1-4614-6141-8_7.

Pliny the Elder. *Natural History: A Selection.* Translated with an introduction and notes by John F. Healy. Penguin Classics. New York: Penguin Books, 1991.

Plutarch. *Lives, with an English Translation by Bernadotte Perrin.* Cambridge, Mass.: Harvard University Press, 1916.

———. *Moralia.* Translated by Harold Cherniss and William C. Helmbold. Cambridge, Mass.: Harvard University Press, 1957.

———. *Roman Lives: A Selection of Eight Roman Lives.* Translated by Robin Waterfield. 1999; reissued New York: Oxford University Press, 2009.

Prendergast, Frank. "Techniques of Field Survey." In *Handbook of Archaeoastronomy and Ethnoastronomy,* ed. C. L. N. Ruggles, 389–409. New York: Springer, 2015. https://doi.org/10.1007/978-1-4614-6141-8_31.

Pumfrey, Stephen. "Harriot's Maps of the Moon: New Interpretations." *Notes and Records of the Royal Society of London.* Royal Society. https://doi.org/10.2307/40647255.

Rahman, Shahid, Tony Street, and Hassan Tahiri. *The Unity of Science in the Arabic Tradition.* Dordrecht: Springer Netherlands, 2008.

Riva, M. A., et al. "The Disease of the Moon: The Linguistic and Pathological Evolution of the English Term 'Lunatic.'" *Journal of the History of the Neurosciences* 20, no. 1 (January 7, 2011): 65–73. https://doi.org/10.1080/09647 04X.2010.481101.

Rochberg, Francesca. *The Heavenly Writing: Divination, Horoscopy, and Astron-*

omy in Mesopotamian Culture. Cambridge: Cambridge University Press, 2007.

Roenneberg, Till, et al. "A Marker for the End of Adolescence." *Current Biology* 14, no. 24 (December 29, 2004). https://doi.org/10.1016/J.CUB.2004.11.039.

Rose, Sarah. *D-Day Girls: The Spies Who Armed the Resistance, Sabotaged the Nazis, and Helped Win World War II.* 2019; reprint New York: Crown, 2020.

Rovelli, Carlo. *The First Scientist: Anaximander and His Legacy.* Translated by Marion Lignana Rosenberg. Yardley, Pa.: Westholme Publishing, 2011.

Rowland, Wade. *Galileo's Mistake: A New Look at the Epic Confrontation Between Galileo and the Church.* New York: Arcade, 2012.

Ruggles, Clive L. N. "Analyzing Orientations." In *Handbook of Archaeoastronomy and Ethnoastronomy,* ed. C. L. N. Ruggles, 411–25. New York: Springer, 2015. https://doi.org/10.1007/978-1-4614-6141-8_26.

———. "Basic Concepts of Positional Astronomy." In *Handbook of Archaeoastronomy and Ethnoastronomy,* ed. C. L. N. Ruggles, 459–72. New York: Springer, 2015. https://doi.org/10.1007/978-1-4614-6141-8_33.

———. "Best Practice for Evaluating the Astronomical Significance of Archaeological Sites." In *Handbook of Archaeoastronomy and Ethnoastronomy,* ed. C. L. N. Ruggles, 373–88. New York: Springer, 2015. https://doi.org/10.1007/978-1-4614-6141-8_25.

———. "Calendars and Astronomy." In *Handbook of Archaeoastronomy and Ethnoastronomy,* ed. C. L. N. Ruggles, 15–30. New York: Springer, 2015. https://doi.org/10.1007/978-1-4614-6141-8_4.

———. ed. *Handbook of Archaeoastronomy and Ethnoastronomy.* New York: Springer, 2015. https://doi.org/10.1007/978-1-4614-6141-8.

———. "Long-Term Changes in the Appearance of the Sky." In *Handbook of Archaeoastronomy and Ethnoastronomy,* ed. C. L. N. Ruggles, 473–82. New York: Springer, 2015. https://doi.org/10.1007/978-1-4614-6141-8_35.

———. "Nature and Analysis of Material Evidence Relevant to Archaeoastronomy." In *Handbook of Archaeoastronomy and Ethnoastronomy,* ed. C. L. N. Ruggles, 353–72. New York: Springer, 2015. https://doi.org/10.1007/978-1-4614-6141-8_22.

———. "Stellar Alignments—Identification and Analysis." In *Handbook of Archaeoastronomy and Ethnoastronomy,* ed. C. L. N. Ruggles, 517–30. New York: Springer, 2015. https://doi.org/10.1007/978-1-4614-6141-8_39.

Ruggles, Clive L. N., and Gary Urton, eds. *Skywatching in the Ancient World: New Perspectives in Cultural Astronomy.* Boulder: University Press of Colorado, 2007.

"Rummaging Through Earth's Attic for Remains of Ancient Life." *Nature News,* November 2, 2002. https://doi.org/10.1038/news021028-13.

Russell, Bertrand. *A History of Western Philosophy.* 1967; reprint New York: Simon & Schuster/Touchstone, 2007.

Salt, Alun. "Development of Archaeoastronomy in the English-Speaking World." In *Handbook of Archaeoastronomy and Ethnoastronomy,* ed. C. L. N. Ruggles, 213–26. New York: Springer, 2015. https://doi.org/10.1007/978-1-4614-6141-8_17.

Samson, David R., et al. "Does the Moon Influence Sleep in Small-Scale Societies?" *Sleep Health* 4, no. 6 (2018): 509–14. https://doi.org/10.1016/j.sleh.2018.08.004.

Shakespeare, William. *Julius Caesar.* Edited by Barbara A. Mowat and Paul Werstine. New York: Simon & Schuster, 2004.

Sherrod, Robert. *Tarawa: The Incredible Story of One of World War II's Bloodiest Battles.* New York: Skyhorse, 2013.

Shubin, Neil. *Your Inner Fish: A Journey into the 3.5-Billion-Year History of the Human Body.* Reprint ed. New York: Vintage, 2009.

Smith, Holland M., and Percy Finch. *Coral and Brass.* New York: C. Scribner's Sons, 1949.

Sobel, Dava. *Galileo's Daughter.* New York: Penguin Books, 2000.

———. *A More Perfect Heaven: How Copernicus Revolutionized the Cosmos.* New York: Walker Books, 2011.

———. *The Planets.* Reprint ed. New York: Penguin Books, 2006.

Solms, Juerg. "Taste of Amino Acids, Peptides, and Proteins." *Journal of Agricultural and Food Chemistry* 17, no. 4 (July 1, 1969): 686–88. https://doi.org/10.1021/jf60164a016.

Somervill, Barbara A. *Nicolaus Copernicus: Father of Modern Astronomy.* Compass Point Books, 2005.

Sordello, Romain, et al. "A Plea for a Worldwide Development of Dark Infrastructure for Biodiversity—Practical Examples and Ways to Go Forward." *Landscape and Urban Planning* 219 (March 1, 2022): 104332. https://doi.org/10.1016/J.LANDURBPLAN.2021.104332.

Šprajc, Ivan. "Alignments upon Venus (and Other Planets)—Identification and Analysis." In *Handbook of Archaeoastronomy and Ethnoastronomy,* ed. C. L. N. Ruggles, 507–16. New York: Springer, 2015. https://doi.org/10.1007/978-1-4614-6141-8_38.

Spudis, Paul D. *The Once and Future Moon.* Washington, D.C.: Smithsonian Books, 1998.

———. *The Value of the Moon: How to Explore, Live, and Prosper in Space Using the Moon's Resources.* Washington, D.C.: Smithsonian Books, 2022.

Steele, John M. "Astronomy and Politics." In *Handbook of Archaeoastronomy and Ethnoastronomy,* ed. C. L. N. Ruggles, 93–101. New York: Springer, 2015. https://doi.org/10.1007/978-1-4614-6141-8_6.

Stewart, Thomas A., et al. "Fin Ray Patterns at the Fin-to-Limb Transition." *Proceedings of the National Academy of Sciences* 117, no. 3 (December 30, 2019): 1612–20. https://doi.org/10.1073/pnas.1915983117.

Stone, Robert, and Alan Andres. *Chasing the Moon: The People, the Politics, and the Promise That Launched America into the Space Age.* New York: Ballantine Books, 2019.

Strano, Giorgio. "The Marāgha School and Copernicus: Scientific Heritage or Independent Elaboration?" In *Circolazione dei saperi nel Mediterraneo: Filosofia e scienze . . .* Florence: Cadmo, 2012. https://doi.org/10.1400/206551.

Stuhlinger, Ernst, and Michael J. Neufeld. "Wernher von Braun and Concentration Camp Labor: An Exchange." *German Studies Review* 26, no. 1 (2003): 121–26. https://doi.org/10.2307/1432905.

Suetonius. *The Twelve Caesars.* Translated by Robert Graves. London: Penguin Books, 1989.

Sullivan, S., P. Mažeika, Katie Hossler, and Lars A. Meyer. "Artificial Lighting at Night Alters Aquatic-Riparian Invertebrate Food Webs." *Ecological Applications* 29, no. 1 (January 2019): e01821. https://doi.org/10.1002/eap.1821.

Thomson, William. *Tides and the Ocean: Water's Movement Around the World, from Waves to Whirlpools.* New York: Black Dog & Leventhal, 2018.

Tidau, Svenja, et al. "Marine Artificial Light at Night: An Empirical and Technical Guide." *Methods in Ecology and Evolution* 12, no. 9 (September 1, 2021): 1588–601. https://doi.org/10.1111/2041-210X.13653.

Timberlake, Todd, and Paul Wallace. *Finding Our Place in the Solar System: The Scientific Story of the Copernican Revolution.* Cambridge: Cambridge University Press, 2019.

Verne, Jules. *From the Earth to the Moon.* Translated by Lowell Bair. New York: Bantam Classic, 1993.

Vogt, David. "Medicine Wheels of the Great Plains." In *Handbook of Archaeoastronomy and Ethnoastronomy,* ed. C. L. N. Ruggles, 541–50. New York: Springer, 2015. https://doi.org/10.1007/978-1-4614-6141-8_41.

Vyazovskiy, Vladyslav V., and Russell G. Foster. "Sleep: A Biological Stimulus from Our Nearest Celestial Neighbor?" *Current Biology* 24, no. 12 (June 16, 2014): R557–60. https://doi.org/10.1016/J.CUB.2014.05.027.

Waltham, David. "Is Earth Special?," *Earth-Science Reviews* 192 (May 1, 2019): 445–70. https://doi.org/10.1016/J.EARSCIREV.2019.02.008.

Ward, Bob, and John Glenn. *Dr. Space: The Life of Wernher von Braun.* Reissue ed. Annapolis, Md.: Naval Institute Press, 2009.

Ward, Peter D., and Donald Brownlee. *Rare Earth: Why Complex Life Is Uncommon in the Universe.* New York: Copernicus, 2003.

White, Jonathan. *Tides: The Science and Spirit of the Ocean.* San Antonio, Tex.: Trinity University Press, 2017.

Williamson, Ray. "Pueblo Ethnoastronomy." In *Handbook of Archaeoastronomy and Ethnoastronomy,* ed. C. L. N. Ruggles, 641–48. New York: Springer, 2015. https://doi.org/10.1007/978-1-4614-6141-8_248.

———. "Sun-Dagger Sites." In *Handbook of Archaeoastronomy and Ethnoastronomy,* ed. C. L. N. Ruggles, 621–28. New York: Springer, 2015. https://doi.org /10.1007/978-1-4614-6141-8_247.

Wood, A. "Asymmetry of the Moon," *Abstracts of the Lunar and Planetary Science Conference* 4 (March 1, 1973): 790. https://ui.adsabs.harvard.edu/abs /1973LPI.....4..790W.

Woolley, Leonard. *Excavations at Ur: A Record of Twelve Years' Work.* New York: Thomas Y. Crowell, 1965.

———. *Ur of the Chaldees.* Rev. ed. Ithaca, N.Y.: Cornell University Press, 1982.

Zantke, Juliane, Heinrich Oberlerchner, and Kristin Tessmar-Raible. "Circadian and Circalunar Clock Interactions and the Impact of Light in *Platynereis dumerilii.*" In *Annual, Lunar, and Tidal Clocks: Patterns and Mechanisms of Nature's Enigmatic Rhythms,* ed. Hideharu Numata and Barbara Helm, 143–62. Tokyo: Springer Japan, 2014. https://doi.org/10.1007/978-4-431-55261-1_8.

Zettler, Richard L., and Lee Horne, eds. *Treasures from the Royal Tombs of Ur.* Philadelphia: University of Pennsylvania Museum of Archaeology and Anthropology, 1998.

Zotti, Georg. "Visualization Tools and Techniques." In *Handbook of Archaeoastronomy and Ethnoastronomy,* ed. C. L. N. Ruggles, 445–57. New York: Springer, 2015. https://doi.org/10.1007/978-1-4614-6141-8_32.

For additional references, please see the author's website at
rebeccaboyle.com/moon.

NOTES

Introduction

1. Sherrod, *Tarawa*, 42.
2. Alexander, *Across the Reef*, 8.
3. The term "spring tide" has nothing to do with the season, but rather the idea of the tide "springing forth." The term "neap tide" comes from the Middle English word *neep*, which meant "scant" or "lacking."
4. Ambrose, *D-Day*, 71–73.
5. Eisenhower, *Crusade in Europe*, 239.
6. Churchill, *Second World War*, 591; and Olson, *Celestial Sleuth*, 254.
7. Gavin, *Aerial Warfare*, 57.

Part I: How the Moon Was Made

Chapter One: A World Apart

1. Collins, *Carrying the Fire*, 393.
2. Aldrin and Abraham, *Magnificent Desolation*, 34.
3. Collins, *Carrying the Fire*, 392.
4. "Vacuuming Equipment," vol. 2, sec. 12.41, *The Apollo 11 Technical Crew Debriefing*, July 31, 1969, https://history.nasa.gov/alsj/a11/a11tcdb.html#120.
5. Hansen, *First Man*, 532.
6. Harrison "Jack" Schmitt is a regular attendee at the Lunar and Planetary Science Conference, an annual meeting of selenographers and geologists, where he frequently answers questions and contributes to scientific pub-

lications. He likened the Moon's smell to gunpowder at a meeting I attended in 2019.

7. David Schleicher, ed., *Interagency Report: Astrogeology 21; Paraphrased Geologic Excerpts from Apollo 12 Mission*, Geological Survey for the National Aeronautics and Space Administration, June 1970, https://www.lpi .usra.edu/resources/USGS-Reports/Astro-0021.pdf.

8. *Apollo 17 Technical Air-to-Ground Voice Transcription* (Houston: Manned Spacecraft Center, 1972), Tape 94A/2, EVA 3, December 13, 1972, 1126–27, https://www.hq.nasa.gov/alsj/a17/AS17_TEC.PDF. See also Rebecca Boyle, "Apollo-Era Tremors Reveal a Dynamic, Active Moon," *Scientific American,* May 13, 2019.

9. Thomas R. Watters et al., "Shallow Seismic Activity and Young Thrust Faults on the Moon," *Nature Geoscience* 12 (2019): 411–17, https://doi .org/10.1038/s41561-019-0362-2.

10. Yosio Nakamura, Gary V. Latham, and H. James Dorman, "Apollo Lunar Seismic Experiment—Final Summary," *Journal of Geophysical Research* 87 (1982): A117–23, https://doi.org/10.1029/jb087is01p0a117.

11. J. P. Williams et al., "The Global Surface Temperatures of the Moon as Measured by the Diviner Lunar Radiometer Experiment," *Icarus* 283 (2017): 300–25, https://doi.org/10.1016/j.icarus.2016.08.012.

12. Laurence R. Harris et al., "How Much Gravity Is Needed to Establish the Perceptual Upright?" *PLoS One* 9 (2014): e106207, https://doi.org/10.1371 /journal.pone.0106207.

13. Philip T. Metzger, "Dust Transport and Its Effects Due to Landing Spacecraft," in *The Impact of Lunar Dust on Human Exploration,* ed. Joel S. Levine (Cambridge Scholars Publishing, 2021).

14. Collins, *Carrying the Fire,* 402.

Chapter Two: The Creation

1. Dalley, *Myths from Mesopotamia*, 233–55.

2. Angular momentum is a quantity that describes the movement and mass of a rotating object or a system of rotating objects: the spinning Earth, the spinning Moon revolving around the spinning Earth, and so forth. The rotation of an object remains constant unless it is acted upon by an external force. Plato talks about this paradoxical pairing of rest and motion in his *Republic.* Sir Isaac Newton was the first Western thinker to lay out the mathematics behind angular momentum, in his *Principia.* The

proof for conservation of angular momentum is attributed to both Daniel Bernoulli and Leonhard Euler, who described it in 1746.

3. Darwin, *Tides and Kindred Phenomena*, 282.

4. Daly, "Origin of the Moon," 104–19.

5. Kant, *Universal Natural History*, pt. 1, sec. 1, 46.

6. U. Wiechert et al., "Oxygen Isotopes and the Moon-Forming Giant Impact," *Science* 294, no. 5541 (2001): 345–48, https://doi.org/10.1126/science.106303.

7. Matija Ćuk and Sarah T. Stewart, "Making the Moon from a Fast-Spinning Earth: A Giant Impact Followed by Resonant Despinning," *Science* 338, no. 6110 (2012): 1047–52, https://doi.org/10.1126/science.1225542.

8. J. A. Kegerreis et al., "Immediate Origin of the Moon as a Post-impact Satellite," *Astrophysical Journal Letters* 937, no. 2 (2022), https://doi.org/10.3847/2041-8213/ac8d96.

9. Mott T. Greene, "Alfred Wegener and the Origin of Lunar Craters," *Earth Sciences History* 17, no. 2 (1998): 111–38, https://www.jstor.org/stable/24138636.

10. Alfred Wegener, "The Origin of Lunar Craters," trans. A. M. Celâl Şengör, *Moon* 14 (October 1975): 211–36, https://doi.org/10.1007/BF00565323.

11. Qian Yuan, "Giant Impact Origin for the Large Low Shear Velocity Provinces," paper presented at the 52nd Lunar and Planetary Science Conference, March 2021, Houston, Tex.

12. Jialong Lai et al., "First Look by the Yutu-2 Rover at the Deep Subsurface Structure at the Lunar Farside," *Nature Communications* 11, no. 1 (2020): 3426, https://doi.org/10.1038/s41467-020-17262-w.

Chapter Three: The Biographer of Earth

1. Cullen, *Heavenly Numbers*, 152.

2. Plato, *Timaeus*, in *Timaeus and Critias*, trans. Desmond Lee (London: Penguin Books, 1965), 54.

3. Ibid.

4. J. Laskar, F. Joutel, and P. Robutel, "Stabilization of the Earth's Obliquity by the Moon," *Nature* 361 (1993): 615–17, https://doi.org/10.1038/361615a0.

5. V. Gysembergh, P. J. Williams, and E. Zingg, "New Evidence for Hipparchus' Star Catalogue Revealed by Multispectral Imaging," *Journal for the History of Astronomy* 53, no. 4 (2022): 383–93, https://doi.org/10.1177/00218286221128289.

6. David Waltham, "Testing Anthropic Selection: A Climate Change Example," *Astrobiology* 11 (2011): 105–14, https://doi.org/10.1089/ast.2010.0475.

7. Christoph Sens-Schönfelder and Tom Eulenfeld, "Probing the *in situ* Elastic Nonlinearity of Rocks with Earth Tides and Seismic Noise," *Physical Review Letters* 122 (2019): 138501, https://doi.org/10.1103/PhysRevLett.122.138501.

8. Denis Andrault et al., "The Deep Earth May Not Be Cooling Down," *Earth and Planetary Science Letters* 443 (2016): 195–203, https://doi.org/10.1016/j.epsl.2016.03.020.

9. Fouad Tera et al., "A Lunar Cataclysm at ~3.95AE and the Structure of the Lunar Crust," in *Lunar Science IV: Abstracts of Papers Presented at the Fourth Lunar Science Conference,* ed. J. W. Chamberlain and Carolyn Watkins (Houston: Lunar Science Institute, 1972), 725.

10. Emerson Speyerer et al., "Quantifying Crater Production and Regolith Overturn on the Moon with Temporal Imaging," *Nature* 538 (2016): 215–18, https://doi.org/10.1038/nature19829.

11. Prabal Saxena et al., "Was the Sun a Slow Rotator? Sodium and Potassium Constraints from the Lunar Regolith," *Astrophysical Journal Letters* 876 (2019): L16, https://doi.org/10.3847/2041-8213/ab18fb.

12. Alex Teachey and David Kipping, "Evidence for a Large Exomoon Orbiting Kepler-1625b," *Science Advances* 5 (2018): 10, https://doi.org/10.1126/sciadv.aav1784.

13. David Ferreira et al., "Climate at High-Obliquity," *Icarus* 243 (November 15, 2014): 236–48, https://doi.org/10.1016/J.ICARUS.2014.09.015.

Part II: How the Moon Made Us

Chapter Four: The Moon and the Origin of Species

1. In 2022 the paleontologists who discovered *Tiktaalik* found a relative, dubbed *Qikiqtania wakei,* that left the land life behind and went back to living in open water. Its paddle-like upper arm seems to have evolved from earlier forms that walked on land, indicating that its more recent ancestors had returned to the sea. See T. A. Stewart et al., "A New Elpistostegalian from the Late Devonian of the Canadian Arctic," *Nature* 608 (2022): 563–68, https://doi.org/10.1038/s41586-022-04990-w.

2. A day in the Lower Devonian, four hundred million years ago, lasted 21.4 hours. By the Upper Devonian and the beginning of land-based vertebrates, the day had lengthened to 22 hours.

3. L. Sallan et al., "The Nearshore Cradle of Early Vertebrate Diversification," *Science* 362, no. 6413 (2018): 460–64, https://doi.org/10.1126/science.aar3689.

4. Steven A. Balbus, "Dynamical, Biological and Anthropic Consequences of

Equal Lunar and Solar Angular Radii," *Proceedings of the Royal Society A* 470, no. 2168 (2014), https://doi.org/10.1098/rspa.2014.0263.

5. H. M. Byrne et al., "Tides: A Key Environmental Driver of Osteichthyan Evolution and the Fish-Tetrapod Transition?" *Proceedings of the Royal Society A* 476 (2020): 20200355, https://doi.org/10.1098/rspa.2020.0355.

6. P. W. Barlow, "Leaf Movements and Their Relationship with the Lunisolar Gravitational Force," *Annals of Botany* 116, no. 2 (2015): 149–87, https://doi.org/10.1093/aob/mcv096.

7. Aristotle, *On the Parts of Animals,* trans. William Ogle (Oxford: Clarendon Press, 1911), https://wellcomecollection.org/works/swtts8b7.

8. Che-Hung Lin et al., "Moonrise Timing Is Key for Synchronized Spawning in Coral *Dipsastraea speciosa,*" *Proceedings of the National Academy of Sciences* 118, no. 34 (2021): e2101985118, https://doi.org/10.1073/pnas.2101985118.

9. Lin Zhang et al., "Dissociation of Circadian and Circatidal Timekeeping in the Marine Crustacean *Eurydice pulchra,*" *Current Biology* 23, no. 19 (2013): 1863–73, https://doi.org/10.1016/j.cub.2013.08.038.

10. Luis M. San-Jose et al., "Differential Fitness Effects of Moonlight on Plumage Colour Morphs in Barn Owls," *Nature Ecology and Evolution* 3 (2019): 1331–40, https://doi.org/10.1038/s41559-019-0967-2.

11. Peter W. Barlow et al., "Leaf Movements of Bean Plants and Lunar Gravity," *Plant Signaling and Behavior* 3 (2008), 1083–90, https://doi.org/10.4161/psb.3.12.6906.

12. Joachim Fisahn, Emile Klingelé, and Peter Barlow, "Lunar Gravity Affects Leaf Movement of *Arabidopsis thaliana* in the International Space Station," *Planta* 241 (2015): 1509–18, https://doi.org/10.1007/s00425-015-2280-x.

13. Marshack, *Roots of Civilization,* 45.

14. C. Helfrich-Förster et al., "Women Temporarily Synchronize Their Menstrual Cycles with the Luminance and Gravimetric Cycles of the Moon," *Science Advances* 7, no. 5 (2021), https://doi.org/10.1126/sciadv.abe1358.

15. Michael Rappenglück, "A Palaeolithic Planetarium Underground—The Cave of Lascaux," *Migration and Diffusion* 5, no. 18 (2004): 93–119.

16. David E. Vance, "Belief in Lunar Effects on Human Behavior," *Psychological Reports* 76 (1995): 32–34, https://doi.org/10.2466/pro.1995.76.1.

17. Jillian C. Banfield, Mohamed Abdolell, and Jai S. Shankar, "Secular Pattern of Aneurismal Rupture with the Lunar Cycle and Season," *Interventional Neuroradiology* 23, no. 1 (2017): 60–63. https://doi.org/10.1177/1591019916675632.

18. T. A. Wehr, "Bipolar Mood Cycles Associated with Lunar Entrainment of a Circadian Rhythm," *Translational Psychiatry* 8, no. 151 (2018), https://doi.org/10.1038/s41398-018-0203-x.

19. Christian Cajochen et al., "Evidence That the Lunar Cycle Influences

Human Sleep," *Current Biology* 23 (2013), https://doi.org/10.1016/j.cub
.2013.06.029.

20. Historian of religion Mircea Eliade argues that lunar symbolism, even a
metaphysics of the Moon, allowed humans to relate to a slew of otherwise impenetrable concepts like "birth, becoming, death, and resurrection; the waters, plants, woman, fecundity, and immortality; the cosmic
darkness, prenatal existence, and life after death, followed by a rebirth of
lunar type ('light coming out of darkness'); weaving, the symbol of the
'thread of life,' fate, temporality, and death; and yet others." Eliade, *Sacred
and Profane*, 156.

21. Bennett Bacon et al., "An Upper Palaeolithic Proto-writing System and
Phenological Calendar," *Cambridge Archaeological Journal* (2023): 1–19,
https://doi.org/10.1017/S0959774322000415; cf. Michael A. Rappenglück,
"A Palaeolithic Planetarium Underground."

Chapter Five: The Beginning of Time

1. Gordon S. Maxwell, "Air Photography and the Work of the Royal Commission on the Ancient and Historical Monuments of Scotland," *Aerial Archaeology* 2 (1978): 37–44.

2. Murray, Murray, and Fraser, *Tale of Unknown Unknowns*, 22.

3. M. G. L. Baillie and M. A. R. Munro, "Irish Tree Rings, Santorini and Volcanic Dust Veils," *Nature* 332 (1988): 344–46, https://doi.org/10.1038
/332344a0.

4. M. G. L. Baillie, "Hekla 3: How Big Was It?" *Endeavour* 13, no. 2 (1989):
78–81, https://doi.org/10.1016/0160-9327(89)90006-9.

5. David Keys, "Cloud of Volcanic Dust Blighted Northern Britain 3000 Years
Ago," *Wat on Earth*, August 16, 1988, https://uwaterloo.ca/wat-on-earth/
news/cloud-volcanic-dust-blighted-northern-britain-3000-years-ago.

6. Stone circles broadly echo the movements of the Sun and Moon across
the sky. If you stood at the Easter Aquhorthies recumbent and followed
the circle toward the northeast, you would mirror the Sun's direction
across the sky after the winter solstice, when the Sun grows stronger and
life on Earth experiences a time of birth and renewal. The stone pillar opposite the recumbent stone represents the summer solstice. If you kept
moving back toward the recumbent, still tracing the circle, you would
walk toward the winter solstice as the Sun falls lower in the sky and life
on Earth experiences death and decay. But beyond these meanings, the

stone circles of northeastern Scotland are also uniquely attuned to the periodic cycles of the Moon.

7. Richard Bradley, *The Significance of Monuments: On the Shaping of Human Experience in Neolithic and Bronze Age Europe* (Abingdon, U.K.: Routledge, 1998), 29–30, 90.

8. Bradley, *Moon and Bonfire*, 112.

9. Scholars have long wondered whether Stonehenge, probably the most famous megalithic monument, was erected someplace else, such as Wales, and later moved to its current home in south-central England. According to the historian Geoffrey of Monmouth, who wrote *The History of the Kings of Britain* in 1136 C.E., the great wizard Merlin's Druid army stole the Stonehenge monoliths from a mythical Irish stone circle called the Giants' Dance. In 2020 a British archaeologist and colleagues published evidence that this may have been true and that Stonehenge was erected in Wales, at a site called Waun Mawn, sometime before 3000 B.C.E. But later research in 2022 showed no link between Waun Mawn and Stonehenge. See Mike Parker Pearson et al., "The Original Stonehenge? A Dismantled Stone Circle in the Preseli Hills of West Wales," *Antiquity* 95, no. 379 (2021): 85–103, https://doi.org/10.15184/aqy.2020.239, and R. E. Bevins et al., "Identification of the Source of Dolerites Used at the Waun Mawn Stone Circle in the Mynydd Preseli, West Wales and Implications for the Proposed Link with Stonehenge," *Journal of Archaeological Science: Reports* 45 (2022): 103556, https://doi.org/10.1016/j.jasrep.2022.103556.

Chapter Six: Early Civilization and the Compass of Time

1. A small number of archaeologists remain unconvinced the Nebra sky disk is evidence of an advanced civilization. In September 2020, Rupert Gebhard of the University of Munich and Rüdiger Krause of the University of Frankfurt argued that the sky disk was probably not of the same origin as the hoard with which Renner and Westphal found it. They say its findspot on the Mittelberg shows evidence of an Iron Age fort, but nothing older, nothing dating to the Bronze Age. They disputed interpretations of soil samples where the disk was found, too. Scientists who studied the dirt said it contained trace amounts of gold and copper, suggesting metals from the disk leached into the earth over thousands of years. But Gebhard and Krause were unconvinced that those bits of gold and copper were related to the disk. Meller and colleagues shot back with

detailed data, including court findings. Other scientists dispute Gebhard and Krause's disputation. In an interview, the British archaeologist Alison Sheridan told me the argument was unseemly and chalked the whole thing up to sour grapes. See Rupert Gebhard and Rüdiger Krause, "Critical Comments on the Find Complex of the So-Called Nebra Sky Disk," *Archäologische Informationen* 43 (2020): 325–46.

2. Hermann Hunger and Erica Reiner, "A Scheme for Intercalary Months from Babylonia," *Wiener Zeitschrift für die Kunde des Morgenlandes* 67 (1975): 21–28, https://www.jstor.org/stable/23868339.

3. Hermann Hunger and David Edwin Pingree, *MUL.APIN: An Astronomical Compendium in Cuneiform* (Ann Arbor: University of Michigan Press, 1989), II:ii:5–6. See also Eshbal Ratzon, "Early Mesopotamian Intercalation Schemes and the Sidereal Month," *Mediterranean Archaeology and Archaeometry* 16, no. 4 (2016): 143–51, https://doi.org/10.5281/zenodo.220913.

4. By 1177 B.C.E., large empires and small kingdoms in the Aegean region, Egypt, and the Near East fell apart in dramatic fashion. The Egyptians, Babylonians, Minoans, Mycenaeans, Hittites, and other civilizations plunged into a centuries-long dark age. Historian Eric Cline argues that the simultaneous demise of so many ancient civilizations was brought about by multiple stressors, especially drought and famine that led to marauding invasions. Cline, *1177 B.C.*

5. The deliberate destruction of Warren Field's timber hall is just one example. At Çatalhöyük in Turkey, some of the earliest known dwellings in human history were ritualistically burned and sealed off. The same thing occurred in the Cahokia civilization, remains of which are located east of St. Louis. For a more thorough discussion of these practices, see Newitz, *Four Lost Cities*. See also R. Hadad, "Ruin Dynamics: Architectural Destruction and the Production of Sedentary Space at the Dawn of the Neolithic Revolution," *Journal of Social Archaeology* 19, no. 1 (2019): 3–26, https://doi.org/10.1177/1469605318794241.

Chapter Seven: The Ornament of the Sky

1. This translation is found in Oates, *Babylon,* 132. An alternate translation appears in Beaulieu, *Reign of Nabonidus,* as "I am Nabonidus, the only son, who has nobody. In my mind there was no thought of kingship."

2. "Before Woolley," Ur Online, n.d., http://www.ur-online.org/about/5/.

3. Woolley, *Ur of the Chaldees,* 203.

4. Woolley, *Excavations at Ur,* 115. Woolley and his contemporaries translit-erated the name Ennigaldi-Nanna, written in cuneiform letters, as Bel-Shalti Nannar, but modern scholars have modified the Latinized spelling. She is also commonly just called "Ennigaldi."

5. "The Herds of Nanna," Electronic Text Corpus of Sumerian Literature, https://etcsl.orinst.ox.ac.uk/section4/tr41306.htm.

6. "The Temple Hymns" (c.4.80.1), line c4801.173, Electronic Text Corpus of Sumerian Literature, https://etcsl.orinst.ox.ac.uk/edition2/etcslgloss .php?lookup=c4801.173.

7. "Ur," Open Richly Annotated Cuneiform Corpus, http://oracc.iaas.upenn .edu/epsd2/cbd/sux/00048262.html.

8. Bauer, *History of the Ancient World,* 98.

9. A description of Enheduanna's life and her role in Sargon's court can be found in Meador and Maier, *Princess, Priestess, Poet,* 5, 78, 245.

10. Support for this analysis is found in Kriwaczek, *Babylon,* 120.

11. Ibid., 122.

12. Enheduanna, "A Hymn to Inanna," trans. t.4.07.3, Electronic Text Corpus of Sumerian Literature, https://etcsl.orinst.ox.ac.uk/cgi-bin/etcsl.cgi ?text=t.4.07.3#.

13. Woolley, *Ur of the Chaldees,* 206.

14. 2 Corinthians 5:17 (King James Version).

15. "The Lament for Urim," trans. t.2.2.2, Electronic Text Corpus of Sumerian Literature, https://etcsl.orinst.ox.ac.uk/cgi-bin/etcsl.cgi?text=t.2.2.2#. See also Bauer, *History of the Ancient World,* 143–44.

16. Genesis 17:5, 17:15 (New International Version).

17. Beaulieu, "Nabonidus the Mad King: A Reconsideration of His Steles from Harran and Babylon," in *Representations of Political Power: Case Histories from Times of Change and Dissolving Order in the Ancient Near East* (University Park: Penn State University Press, 2007), 138–140, https://doi.org /10.1515/9781575065830-010.

18. Ibid.

19. Francesca Minen, "Touched by the Moon: Lunar Influences on Human Health in Ancient Mesopotamia," in *La luna nell'immaginario umano,* ed. Lara Nicolini, Luca Beltrami, and Lara Pagani (Genoa: Genoa University Press, 2019), 48.

20. Beaulieu, *Reign of Nabonidus,* 208. Alternate translation in C. J. Gadd, "The Harran Inscriptions of Nabonidus," *Anatolian Studies* 8 (1958): 35–92.

21. Gadd, "Harran Inscriptions of Nabonidus," 47.

22. Ibid.

23. Al Wolters, "Belshazzar's Feast and the Cult of the Moon God Sîn," *Bulletin for Biblical Research* 5 (1995): 199–206, https://www.jstor.org/stable/26422134.

24. Scholars say the feast may have been an Akitu (New Year's) festival in honor of the Moon God Sin. The dates, the sixteenth and seventeenth day of Tišritum, would always have followed the October full Moon, the harvest moon. Other evidence backs this up. A papyrus manuscript from the third century B.C.E. found in Thebes, Egypt, called Papyrus Amherst 63, also describes a Mesopotamian festival dedicated to the Moon God. The papyrus (in the Morgan Library & Museum in New York) is written in Aramaic and contains Israelite psalms not found in the Bible corpus. "Crescent, be a bow in heaven!" one psalm reads. "Our Bull shall be with us." Given the religious politics of Nabonidus's reign, it seems reasonable to guess that a festival worshipping the Moon God Sin, not Marduk, was taking place when Babylon fell. As far as the Moon phase itself, retired astrophysicist and solar eclipse chaser Fred Espenak maintains a searchable record of Moon phases dating to 2000 B.C.E., and shows a full Moon on the night of October 24, 539 B.C.E.

25. William H. Shea, "Nabonidus, Belshazzar, and the Book of Daniel: An Update," *Andrews University Seminary Studies* 20, no. 2 (Summer 1982): 133–49.

26. By 539 B.C.E., so many societies were literate, and Babylon was so powerful and so renowned, that its fall was documented in numerous places. For example, the narrative here comes from the Cyrus Cylinder; the *Verse Account of Nabonidus*, which was written later by the enemies of the Friend of Sin; *Cyropaedia: The Education of Cyrus*, by the Greek historian Xenophon and translated by Henry Graham Dakyns; *The History of Herodotus*, translated by G. C. Macaulay; and the Book of Daniel (New American Bible).

27. Aristotle, *Politics*, vol. 21 of *Aristotle in 23 Volumes*, trans. H. Rackham (Cambridge, Mass.: Harvard University Press, 1944), 3.1276a.

28. Rochberg, *Heavenly Writing*, 78.

29. Ibid., 40.

30. Campion, *Astrology and Popular Religion*, 13. See also the nationally representative YouGov poll, conducted April 21–22, 2022, https://today.yougov.com/topics/lifestyle/survey-results/daily/2022/04/22/5ad3f/1.

Chapter Eight: The Voyage of Discovery Begins with the Moon

1. Plutarch, *On the Apparent Face in the Moon*, in *Moralia*.

2. "The problem is that for all their fecundity in theorizing, the Presocratics

fail to provide any way to test their theories. So in the end every theory is equally plausible and equally unprovable and unproved," as the Brigham Young University philosopher Daniel W. Graham explains in *Science Before Socrates*, 10. Graham has long been a voice in the wilderness extolling the work of pre-Socratic thinkers, especially Anaxagoras. See also Graham, *Explaining the Cosmos*.

3. Plato, *Phaedo*, vol. 1 of *Plato in Twelve Volumes*, trans. Harold North Fowler (Cambridge, Mass.: Harvard University Press, 1966).

4. Graham, *Science Before Socrates*, 91.

5. Plutarch, *Lysander*, in *Lives, with Translation by Perrin*, 12.

6. "Anaxagoras," *Stanford Encyclopedia of Philosophy*, November 11, 2019, https://plato.stanford.edu/entries/anaxagoras/.

7. Translation in Graham, *Science Before Socrates*, 7.

8. Plato, *Euthyphro, Apology, and Crito: With the Death Scene from Phaedo*, trans. F. J. Church, rev. Robert D. Cumming (1956; reprint New York: Macmillan, 1985), 32.

9. Russell, *History of Western Philosophy*, 145.

10. Suetonius, *Twelve Caesars*, 16.

11. Beard, *SPQR*, 285.

12. Fitzpatrick and Haselgrove, *Caesar's Battle for Gaul*, 135–158.

13. Caesar, *Seven Commentaries*, 82.

14. Appian, *Civil Wars*, 78.

15. Ibid., 88.

16. Matthew 22:20–22 (King James Version).

17. Paul Coones, "The Geographical Significance of Plutarch's Dialogue, Concerning the Face Which Appears in the Orb of the Moon," *Transactions of the Institute of British Geographers* 8, no. 3 (1983): 361–72, https://doi.org /10.2307/622050. See also Johannes Kepler, *Somnium, seu Opus posthumum de astronomia lunari*, ed. Ludovico Kepler (Frankfurt, 1634).

18. Plutarch, *On the Apparent Face in the Orb of the Moon*, in *Moralia*, 24:1–3.

Part III: How We Made the Moon

Chapter Nine: The Moon in Our Eyes

1. Quoted in Rowland, *Galileo's Mistake*.

2. Ptolemy does not address these inaccuracies. "This discrepancy is silently ignored by Ptolemy, though he could not have doubted that the actual geocentric distances of the moon were very different from what his

model required," wrote Otto Neugebauer, a historian of science who sets the bar for all others, in *Exact Sciences in Antiquity*, 195.

3. Hassan Tahiri, "The Birth of Scientific Controversies, the Dynamics of the Arabic Tradition and Its Impact on the Development of Science: Ibn al-Haytham's Challenge of Ptolemy's *Almagest*," in Rahman, Street, and Tahiri, *Unity of Science in Arabic Tradition*, 208.

4. Ibid., 209.

5. Moller, *Map of Knowledge*, 227.

6. Nicolaus Copernicus, *Commentariolus*, quoted in Bauer, *Story of Western Science*, 47.

7. Sobel, *More Perfect Heaven*, 50.

8. Nicolaus Copernicus, *De Revolutionibus Orbium Cœlestium, Libri VI*, trans. Edward Rosen (Warsaw: Polish Scientific Publications, 1978), 173. Additional quotes in this section are from this translation.

9. Modern scholars have found evidence that al-Tusi's geometry was translated into both Greek and Hebrew in the thirteenth century, suggesting widespread transmission. The geometrical concept known as the Tusi couple in particular seems to have made its way from Maragha Observatory into Europe two centuries before *Almagest*. This mathematical device describes how a small circle can rotate quickly within a larger circle whose diameter is twice the size of the small one. This arrangement of circles is superior to epicycles in explaining the motions of the planets. Nasir al-Din al-Tusi sketched this concept in 1247, and scholars say his drawing is remarkably similar to one by Copernicus in *De Revolutionibus*. See Willy Hartner, "Copernicus, the Man, the Work, and Its History," *Proceedings of the American Philosophical Society* 117, no. 6 (December 31, 1973): 413–22.

   In 1957, Otto Neugebauer found a Greek-language manuscript in Oxford's Bodleian Library that contains a diagram of the Tusi couple and al-Tusi's lunar model, and this has also been used to suggest that Copernicus might have known of al-Tusi's work. For a detailed overview of this history and how *Almagest* was transmitted to Europe, see Kevin Krisciunas and Belén Bistué, "Notes on the Transmission of Ptolemy's *Almagest* to the Era of Copernicus," *Journal of Astronomical History and Heritage* 22, no. 3 (2019): 492–502. For a discussion of how Copernicus might have learned from Islamic astronomers, see F. Jamil Ragep, "Copernicus and His Islamic Predecessors: Some Historical Remarks," *History of Science* 45, no. 1 (2007): 65–81, https://doi.org/10.1177/007327530704500103.

10. Russell, *History of Western Philosophy*, 582.

11. Quotes from Kepler and from Harriot's contemporaries are in Fox, *Thomas Harriot and His World*, and Arianrhod, *Thomas Harriot*, 213.

12. Arianrhod, *Thomas Harriot*, 146.

13. This is from the first stanza of the poem "High Flight," by John Gillespie Magee Jr., which my pilot father had framed and hung in my childhood home. John Gillespie Magee, "Letter to Parents," September 3, 1941. John Magee Papers, Library of Congress, Washington, D.C. Manuscript. In *Respectfully Quoted: A Dictionary of Quotations Requested from the Congressional Research Service*, ed. Suzy Platt (Washington, D.C.: Library of Congress, 1989): 117–18. Accessed via Poetry Foundation, http://www.poetryfoundation.org/poems/157986/high-fight-627d3cfb1e9b7.

14. Arianrhod, *Thomas Harriot*, 218.

15. These quotes, taken from the margins of Harriot's drawings, were deciphered by the author at Petworth House.

16. For a thorough discussion of the mathematical, cartographic renderings Harriot produced, see Montgomery, *Moon and Western Imagination*, 108.

17. Koestler, *Watershed*, 50.

18. Kepler, *Conversation with Galileo's Sidereal Messenger*, 25.

19. Galilei, *Sidereus nuncius*, 22. Passage translated by the author.

20. Ibid., 20. Passage translated by the author.

21. Kepler, *Conversation with Galileo's Sidereal Messenger*, 18.

22. Galilei, *Dialogue Concerning World Systems*, 77.

23. Ibid. This quotation is part of the same conversation about the "horns" of the crescent Moon.

24. Steven F. Ostrow, "Cigoli's *Immacolata* and Galileo's Moon: Astronomy and the Virgin in Early Seicento Rome," *Art Bulletin* 78, no. 2 (1996): 218–35, https://doi.org/10.2307/3046173. See also Montgomery, *Moon and Western Imagination*, 129.

25. Kepler, *Conversation with Galileo's Sidereal Messenger*, 42.

26. Council of Trent, Fourth Session, *The Canons and Decrees of the Sacred and Ecumenical Council of Trent*, trans. J. Waterworth (London: Dolman, 1848), 17–21.

Chapter Ten: Journeys of the Mind

1. Koestler, *Sleepwalkers*, 343; cf. Kepler's introduction to *Astronomia Nova*, New Revised Edition, trans. William H. Donahue (Santa Fe: Green Lion Press, 2020), 25.

2. Koestler, *Watershed*, 195.

3. Somervill, *Nicolaus Copernicus*, 13.

4. Lear, *Kepler's Dream*, 136.

5. Kepler, *Conversation with Galileo's Sidereal Messenger,* 46. See also Guthke, *Last Frontier,* 108.

6. Koestler, *Watershed,* 39.

7. Marjorie Nicolson, "Kepler, the *Somnium,* and John Donne," *Journal of the History of Ideas* 1, no. 3 (1940): 259–80, https://doi.org/10.2307/2707087.

8. Connor, *Kepler's Witch,* 9.

9. Lear, *Kepler's Dream,* 77.

10. Irving, *Knickerbocker's History,* 14.

11. Ibid.; cf. Irving, *History of New York,* 50.

12. Verne, *From Earth to Moon,* 213.

13. Hurt, *For All Mankind.*

14. Neufeld, *Von Braun,* 4.

15. Several biographies tell von Braun's life story, including his involvement with Nazi politics, use of prisoner labor at concentration camps, and his ultimate success in landing Americans on the Moon. He was a complicated man. For two excellent accounts, see Neufeld, *Von Braun,* and Bergaust, *Wernher von Braun.*

16. Neufeld, *Von Braun,* 5. See also Bergaust, *Wernher von Braun,* 60.

17. McDougall, . . . *Heavens and Earth,* 100.

18. Launius, *Apollo's Legacy,* 160.

19. Randy Liebermann, "The Collier's and Disney Series," in Ordway and Lieberman, *Blueprint for Space,* 146.

Chapter Eleven: The *Eagle* and the Reliquary

1. Borman and Serling, *Countdown,* 212.

2. Muir-Harmony, *Operation Moonglow,* 159.

3. Launius, *Apollo's Legacy,* 215.

4. Muir-Harmony, *Operation Moonglow,* 127.

5. Maher, *Apollo in Age of Aquarius,* 11–12.

6. Barbree and Glenn, *Neil Armstrong,* 156.

7. Ibid., 209.

8. Collins, *Carrying the Fire,* 361.

9. Verne, *From Earth to Moon,* 203.

10. The Apollo 11 landing recordings and accompanying technical transcript are some of the most important records in human history, I would argue. It's worth reading them unvarnished, or even better, listening to them, which I did for the first time when I was about ten years old. This section is taken from the Apollo Technical Air-to-Ground Voice Transmission

transcript, available through NASA's website, at https://www.hq.nasa
.gov/alsj/a11/a11transcript_tec.html.

11. Collins, *Carrying the Fire,* 406.

12. Some of the Apollo 11 transmissions were somewhat garbled, which
should be unsurprising, given the distance the radio waves had to travel
and the limited technology of 1969. Numerous experts have reviewed the
recordings over the years and still argue about transcript writers' interpretation of individual words. Most of the time, this is inconsequential,
but the passage referenced here is one example of a transcription decision that can change the recording's meaning. Aldrin and Armstrong
were out of the lunar lander, and Armstrong read aloud the words on the
lander's plaque, which showed the two hemispheres of Earth and the
words "Here men from the planet Earth first set foot upon the Moon." Aldrin then started talking about the TV camera's lens and worried that it
was smudged with fine Moon dust. He described the dust, called regolith,
as "very finely powdered carbon, but really [garbled] looking." Some transcript writers heard "sooty-looking," while others heard "dirty-looking."
The official edited transcript, which is protected by copyright, has Aldrin
saying "pretty-looking." See Eric M. Jones, "One Small Step," *Apollo 11
Lunar Surface Journal,* https://history.nasa.gov/alsj/a11/a11.step.html.
See also "Apollo 11 Technical Air-to-Ground Transcript," *Apollo 11 Lunar
Surface Journal,* https://www.hq.nasa.gov/alsj/a11/a11trans.html. This is
the raw transcript as it was provided to the news media in 1969.

Another, even better example is the most famous utterance ever made
on the Moon. Some experts still debate whether Neil Armstrong said, or
at least meant to say, "That's one small step for a man," versus "one small
step for man." After returning to Earth, Armstrong said he actually spoke
the words "a man," although most people do not hear it that way. It
makes a difference, because without the indefinite article "a," Armstrong
was using "man" as an abstract reference to all of humanity. It's like saying, "That's one small step for mankind, one giant leap for mankind." But
a generation of listeners still does not hear the "a," and it is almost never
conveyed this way in museum exhibits or, as it happens, in most books. I
hear a slight wobble in his voice as he speaks, but I have never heard the
indefinite article either, no matter how much I wish I could.

13. Jones, "One Small Step."

14. Tony Phillips, "Wide Awake in the Sea of Tranquillity," NASA Science,
July 20, 1969, https://science.nasa.gov/science-news/science-at
-nasa/2006/19jul_seaoftranquillity.

15. Collins, *Carrying the Fire,* 54.

16. This quote is often repeated by evangelical Christians, and comes from Irwin's autobiography. See Irwin, *More Than Earthlings*, 10.

17. Aldrin told this story many times after returning to Earth, including in multiple books and magazine articles. The account here is taken from Aldrin and Abraham, *Magnificent Desolation*, 26.

18. "Day 8, Part 2: More Television and Stowage for Re-entry," *Apollo 11 Flight Journal*, https://history.nasa.gov/afj/ap11fj/25day8-reentry-stowage.html.

19. Among people born in the "Silent Generation," from 1928 to 1945, a whopping 84 percent of Americans identified as Christian, according to the Pew Research Center. See "In U.S., Decline of Christianity Continues at Rapid Pace," Pew Research Center, October 17, 2019, https://www
.pewresearch.org/religion/2019/10/17/in-u-s-decline-of-christianity
-continues-at-rapid-pace/.

20. This is probably one of the most famous, and frameable, quotes from the Apollo era. The best source I can find is from a 1974 feature in *People* magazine, though it is not clear whether Mitchell gave this quote during an interview, or whether the magazine was reprinting it. See "Edgar Mitchell's Strange Voyage," *People*, April 4, 1974, https://people.com
/archive/edgar-mitchells-strange-voyage-vol-1-no-6/.

21. Eric M. Jones, "The Genesis Rock," *Apollo 15 Lunar Surface Journal*, https://
www.hq.nasa.gov/alsj/a15/a15.spur.html. See also Rebecca Boyle, "To See the Moon Anew in Its Primordial Crust," *New York Times*, July 14, 2019.

Chapter Twelve: Our Eighth Continent

1. Ecclesiastes 1:9–10 (King James Version).

2. Martin Elvis, "The Peaks of Eternal Light: A Near-Term Property Issue on the Moon," *Space Policy* 38 (2016): 30–38.

3. Vera Assis Fernandes, "Ethical and Social Aspects of a Return to the Moon—A Geological Perspective," *Geosciences* 9, no. 1 (2019): 12, https://
doi.org/10.3390/geosciences9010012.

4. Enric Volante, "Navajos Upset After Ashes Sent to Moon; Nasa Apologizes," *Arizona Daily Star*, January 15, 1998, https://www.spokesman.com
/stories/1998/jan/15/navajos-upset-after-ashes-sent-to-moon-nasa/. See also Virgiliu Pop, "Lunar Exploration and the Social Dimension," in *Earth-Like Planets and Moons: Proceedings of the 36th ESLAB Symposium*, June 3–8, 2002, ESTEC, Noordwijk, Netherlands, 299–301.

ABOUT THE AUTHOR

Rebecca Boyle is a contributing editor at *Scientific American*, a contributing writer for *Quanta Magazine* and *The Atlantic*, and a frequent contributor at the *New York Times, Popular Science,* Smithsonian's *Air & Space Magazine,* and many other publications. She is a member of the group science blog *The Last Word on Nothing*. Boyle was a Knight Science Journalism Fellow at the Massachusetts Institute of Technology, and is the recipient of numerous writing awards throughout her career. Her work has been anthologized three times in *The Best American Science and Nature Writing*. Boyle is a former newspaper reporter, a former Space Camp attendee, and a lifelong Moon enthusiast. She lives on a mountain in Colorado with her husband, their two daughters, their golden retriever, and several neighboring black bears. This is her first book.

rebeccaboyle.com
Instagram: @by.rebecca.boyle
Twitter: @rboyle31